Astrophysics of Brown Dwarfs

Astrophysics of Brown Dwarfs

Proceedings of a Workshop held at George Mason University
Fairfax, Virginia, October 14–15 1985

Edited by

MINAS C. KAFATOS
George Mason University
Fairfax, Virginia

ROBERT S. HARRINGTON
United States Naval Observatory
Washington, DC

STEPHEN P. MARAN
NASA–Goddard Space Flight Center
Maryland

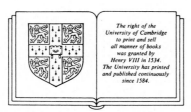

The right of the
University of Cambridge
to print and sell
all manner of books
was granted by
Henry VIII in 1534.
The University has printed
and published continuously
since 1584.

CAMBRIDGE UNIVERSITY PRESS

Cambridge

London   New York   New Rochelle

Melbourne   Sydney

Published by the Press Syndicate of the University of Cambridge
The Pitt Building, Trumpington Street, Cambridge CB2 1RP
32 East 57th Street, New York, NY 10022, USA
10 Stamford Road, Oakleigh, Melbourne 3166, Australia

First Published 1986

Printed in Great Britain at the University Press, Cambridge

Library of Congress cataloguing in publication data available

British Library cataloguing in publication data
Astrophysics of brown dwarfs: proceedings
of a workshop held at George Mason University,
Fairfax, Virginia, October 14–15, 1985.

1. Brown Dwarfs
I. Kafatos, Minas C.   II. Harrington, Robert S.
III. Maran, Stephen P.
523.8'2   QB843.B7/

ISBN 0 521 32337 1

CONTENTS

FOREWORD

This volume includes all the scientific papers presented at the Astrophysics of Brown Dwarfs Workshop, held at George Mason University, Fairfax, Virginia, on October 14-15, 1985. The papers are arranged in the actual order they were presented and the volume is divided into two sections, Observations and Theory. The Editors wrote a brief introduction to each section but the reader is encouraged to examine, perhaps before reading the book in detail, the excellent conference summary at the end of the volume by Dr. John N. Bahcall. We can say that most of the experts in the field of brown dwarfs were present at the meeting, as well as a large number of other astronomers. This probably shows the interest that the identification of VB8 B as a brown dwarf has sparked in the minds of many astronomers. I don't consider myself as an expert on brown dwarfs but I learned a lot of new and exciting astronomy at the workshop.

This is the second George Mason Fall Workshop in Astrophysics. These workshops are held annually in the month of October and for each year a topic in the forefront of astrophysics is chosen by the Organizing Committee. Astronomers are urged to provide their suggestions on future topics since these workshops are intended to serve the astronomical community. I would personally like to thank all the participants, all who presented review and contributed papers and finally Drs. R.S. Harrington and S.P. Maran who provided invaluable help towards the realization of the workshop and its proceedings.

Minas Kafatos
Department of Physics
George Mason University

February 1986

ACKNOWLEDGEMENTS

We would like to thank the other members of the Organizing Committee Drs. L.W. Fredrick, S. Mitton and V.L. Trimble for their assistance and suggestions during the planning stages of this workshop, and Dr. J.C. Evans for his assistance during the workshop. We also thank numerous people at George Mason University, including Drs. George W. Johnson, President of George Mason, David J. King, Vice President for Academic Affairs, William P. Snavely, Dean of the College of Arts and Sciences, Averett S. Tombes, Dean of the Graduate School, and Robert Ehrlich, Chairman of the Physics Department for their generous support of the meeting; Dr. Simon Mitton, Science Editorial Director, Cambridge University Press for generous support of the meeting. Dr. Nancy Joyner, Ms. Mary Betz and the other staff members of the Office of Community Services for their invaluable help in organizing the workshop and assistance during the workshop. We would also like to thank Ms. Diana F. Smith of the Physics Department for her assistance in typing segments of the present manuscript. One of us, M.Kafatos, would also like to thank Dr. F. D'Antona for her seminar and public talk at George Mason University following the end of the workshop.

R.S. Harrington, Co-Chairman
M. Kafatos, Chairman
S.P. Maran, Co-Chairman

WORKSHOP PARTICIPANTS

J.N. Bahcall

W.A. Baum

D.C. Black

P.C. Boeshaar

A. Boss

A. Burrows

B. Campbell

O. Dahari

C. Dahn

F. D'Antona

D.S. Dearborn

D. Deming

H. DeWitt

B. Dorman

R. Ehrlich

J.C. Evans

L.W. Fredrick

A. Fresneau

G.D. Gatewood

R.S. Harrington

M.R.S. Hawkins

R.C. Henry

C.F. Hoeffer

R.S. Hawkins

W.B. Hubbard

P. Hut

P.A. Ianna

M. Kafatos

C.K. Kumar

W. Lewis

J.W. Liebert

V. Lindsay

F.J. Low

J. Lunine

G.W. Marcy

T. Matsumoto

T. Mazeh

C. Max

D.W. McCarthy, Jr.

L. Nelson

F. Palla

S. Perlmutter

P. Pesch

J.L. Phillips

R.G. Probst

S. Rappaport

N.G. Roman

J.L. Russell

D. Saumon

D. Schroeder

H.L. Shipman

M. Skrutskie

S. Stahler

D.J. Stevenson

K.A. Strand

G.S. Stringfellow

B. Tarter

J. Tarter

D.C. Teplitz

V.L. Teplitz

V. Trimble

H. Van Horn

F.R. West

C.E. Worley

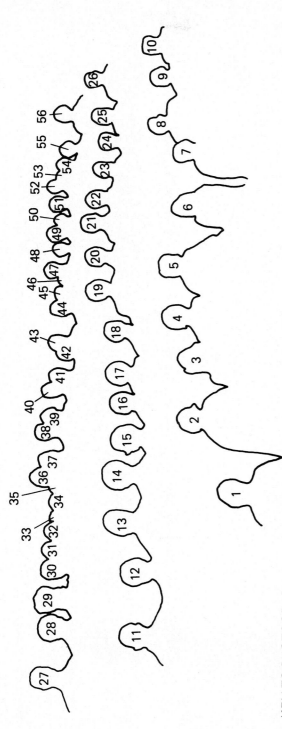

KEY TO PHOTOGRAPH

1. M.R.S. Hawkins	15. F. D'Antona	29. D.S. Dearborn	43. D.J. Stevenson
2. R.G. Probst	16. F.R. West	30. V. Lindsay	44. J.W. Liebert
3. P.C. Boeshaar	17. S.P. Maran	31. T. Mazeh	45. H.L. Shipman
4. A. Fresneau	18. D.C. Black	32. S. Perlmutter	46. C.E. Worley
5. R.C. Henry	19. G.W. Marcy	33. L.W. Fredrick	47. D. Saumon
6. V. Trimble	20. B. Campbell	34. O. Dahari	48. B. Tarter
7. B. Dorman	21. C.C. Dahn	35. T. Matsumoto	49. M. Skrutskie
8. C. Hoeffer	22. D.W. McCarthy	36. A. Boss	50. S. Rappaport
9. M. Kafatos	23. P.A. Ianna	37. G.S. Stringfellow	51. D. Deming
10. J.C. Evans	24. K.A. Strand	38. R.S. Harrington	52. A. Burrows
11. J.N. Bahcall	25. H. DeWitt	39. J. Lunine	53. L. Nelson
12. C. Max	26. W.B. Hubbard	40. J.L. Phillips	54. W. Lewis
13. P. Hut	27. W.A. Baum	41. J. Tarter	55. V.L. Teplitz
14. S. Stahler	28. H. Zinnecker	42. J.L. Russell	56. F. Palla

THE WORKSHOP

I. OBSERVATIONS

Efforts towards the observational detection of Brown Dwarfs are a recent development, owing to the intrinsic faintness of these objects and thus the inherent difficulty in their detection. Classically, the possibility of the existence of a body of this type has been raised by astrometry, but a direct detection is required for confirmation. The best technique for this is speckle interferometry, and it has yielded one definite object (VB8B) so far. Photometric surveys of obviously red, apparently nearby stars have yielded a few candidate objects, but a more thorough understanding of the observational properties of very low-mass objects is required.

A variety of new approaches is being brought to bear on the problem of the observational detection of these objects. High-precision radial velocity monitoring has yielded nothing definite so far, although there is one hopeful possibility. IRAS should have turned up many such objects, but it also has yielded nothing, either as field stars or as companions to white dwarfs. Neither a deep CCD survey in the southern hemisphere nor an IR survey for companions to nearby stars has produced positive results. The systematic search for a very nearby Solar companion is only just getting under way. Only a photographic Schmidt surevy in the South has produced any suggestion of the existence of these objects in any numbers. The hope now is back to astrometry, and the new long-range, high-precision efforts towards detection, both from the ground and from space, will return to this classic approach.

THE ASTROMETRIC DETECTION OF BROWN DWARFS

R. S. Harrington
U. S. Naval Observatory, Washington, DC 20390

Keywords: Brown Dwarfs, Astrometry

ABSTRACT. Brown dwarfs are detected astrometrically from
the systematic variation, or perturbation, from the
expected proper-plus-parallactic motion of a star across
the sky. All such objects detected to date have come from
conventional parallax programs employing long-focus
astrometric telescopes. Using the observed period and
amplitude of the perturbation, plus photometric data and an
assumed mass-luminosity law, it is possible to deduce two
possible values for the mass of the object causing the
effect. If it is possible to rule out a non-brown-dwarf
mass, or it can be demonstarted that the mass-luminosity
law is not applicable, the case for there being a brown
dwarf becomes strong. There are eight cases now known
where there are probable brown dwarfs.

The astrometric detection of brown dwarfs, or any unseen companion for
that matter, is a classic case of a serendipitious discovery in
astronomy. That is, all such detections to date have been by-products
of the astrometric determination of the parallaxes and proper motions
of nearby stars, although the Sproul program has been designed with the
possibility of such detections in mind. There are several programs
being planned to primarily look for faint, unseen companions, but it
will be some time before those efforts can begin to produce results.

Parallax and proper motion programs consist of obtaining many images
over many years of relatively small fields around the stars of interest.
These images are obtained with special long-focus instruments,
traditionally refractors, but more so now reflectors. These images
have also traditionally been obtained photographically, although there
are programs now to use more sensitive electronic detectors. Positions
of the target star are obtained relative to a background reference
frame of presumably very distant stars. From these positions, the
barycentric relative parallax, the relative proper motion, and
sometimes the perspective acceleration can be obtained.

After the conventional astrometric parameters have been derived, the
positional residuals are analyzed for any systematic trends that may
remain. This can be done by straightforward visual inspection of data
or plots, or by more sophisticated numerical procedures. If any such
trends that are detected can also be established as periodic, the
existence of an unseen companion can be surmised.

The parameters of the periodic trend, called a perturbation, that need
to be established are the period, P, and the semi-major axis of the
motion of the photocenter about the barycenter, α. Of course, to obtain
a good value for P, an interval of time considerably longer than the
period must be covered by the observations. If we define a to be the
semi-major axis of the secondary about the primary, B to be the mass
of the secondary compared to the mass of the system, and β to be the
luminosity of the secondary compared to the luminosity of the system,
the following relationship has been established emperically:

$$\alpha/a = B - \beta \tag{1}$$

That is, the photocentric image is pulled back from the primary towards
the barycenter by an amount equal to the light ratio. We also have
Kepler's third law, expressed as follows, where π is the absolute
parallax (obtained from the relative solution with a small statistical
correction);

$$(a/\pi)^3/P^2 = m_1 + m_2 \tag{2}$$

where m_1 is the mass of the primary and m_2 is that of the secondary.

Most good astrometric programs either observe stars with good photoelectric photometry available or include a photometric program of
their own. From the observed visual magnitude of the system and the
parallax, it is possible to obtain the total visual absolute magnitude
of the system. For blended images, this can be resolved into the
individual absolute magnitudes if the relative luminosities are known.
From this, in turn, masses of the stellar components can be determined,
using an emperical mass-visual luminosity relationship.

Armed with all of the above, it is possible to try a wide range of mass
of the secondary, from negligible to equal to the primary, and obtain
possible values of α^3/P^2, and thus the values of the secondary mass
consistent with the observed quantities. Generally, there are two
such solutions, and they are characterized by their limiting cases.
A very small perturbation can mean that the two components are
comparable in both mass and luminosity, and hence both B and β are
close to 0.5, with their difference being small. On the other hand,
the secondary may be practically negligible compare to the primary,
and both B and β are small. There then remains the question of judging
whether one solution is not permitted for other reasons.

Two other situations may exist. One is that the two cases are close
enough that they agree within their errors. The other is that there
is no solution, which usually means the secondary does not obey the
mass-luminosity relationship, i.e. it is substellar, a brown dwarf.
The luminosity ratio is then set identically to zero, and the mass
of the secondary determined now in a very straightforward manner.

Eight systems with reliable perturbations (thus excluding Barnard's and Van Biesbroeck's Stars) have been identified in which the secondary would be a brown dwarf if the negligible mass case proves to be correct. Here, to be conservative, an object must have a mass equal to or less than 0.05 Suns to be classified a potential brown dwarf. Table I lists those systems, in increasing order of mass of the brown dwarf if in fact that is what the companion is. Given are the star designation, the parallax, the perturbation amplitude, the period, the masses of the primary and secondary for each case, and the speckle status (N means negative result, D means positive detection).

Table 1

Star	π	α	P	Comparable m_1	m_2	Negligible m_1	m_2	Status
+68°946	0".213	0".033	26^y.4	0.28	0.27	0.33	0.008	N
+43°4305	0.200	0.039	11.7	0.24	0.22	0.27	0.009	N
L1750-5	0.109	0.013	3.5	0.16	0.14	0.17	0.016	
S2051A	0.181	0.070	23.0	0.20	0.18	0.24	0.018	N
VB8	0.154		decades			0.08	0.02	D
CC1228	0.082	0.012	6.3	0.32	0.29	0.36	0.02	N
G139-29	0.075	0.056	10.0	0.14	0.09	0.14	0.04	
G107-69	0.091	0.013	0.9	0.17	0.12	0.18	0.05	

Looking at these systems in more detail, +68°946 was discussed by Lippincott (1977) based on over 38 years of Sproul data. The comparable mass case would require almost no magnitude difference, and a separation greater than 1.5 arc seconds at times, which should have produce detectably elongated images. Having seen none, the alternate interpretation is preferred. Similarly, +43°4305, discussed by van de Kamp and Worth (1972) based on 35 years of Sproul material, would also have a negligible magnitude difference and over 1.5 arc seconds separations if the components were comparable. Hence, the negligible mass case is preferred here as well.

L1750-5 was first suggested by Dahn et al (1972) as having a companion, although no interpretation was attempted at that time. By now we have 15 years of material on this system, and a preliminary solution for a perturbation (i.e. assuming circular motion) was carried out, with the results presented. We have no reason to prefer either case and thus are awaiting speckle results.

Stein 2051A is the brighter component of the red dwarf/white dwarf pair Stein 2051, and it was discussed by Strand (1977) from 11 years of both Sproul and USNO data. This system is of particular interest, since there is enough curvature in the binary motion to permit a determination of the mass ratio, and hence the mass of the white dwarf if the mass of the red dwarf is known. The comparable component case for the companion of the red dwarf suggests a small magnitude difference and a separation over 1 arc second at times, which should have been detected. This, plus the negative speckle result, argues for the negligible mass unseen companion and the larger red dwarf mass. This system is still under active scrutiny.

VB8 is part of the by now septuple system that includes the binary, Wolf 629, and the triple, Wolf 630. Harrington et al (1983) first suggested the presence of an unseen companion, although it was obvious that only part of the orbit had been covered. The direct imaging of the companion by McCarthy (1985) permitted the imposition of certain limits on the mass of the companion by Harrington (1985). The mass of the companion quoted in the Table is only a most probably value in the loosest sense, and any value in the range 0.01-0.08 is acceptable. As McCarthy has fully demonstrated, this is the object in the Table that is most obviously a true brown dwarf.

CC1228 was discussed by Lippincott (1979) using 40 years of Sproul material. The negative speckle result is the primary argument in favor of a brown dwarf companion. G139-29 was discussed by Christy (1978) from 12 years of USNO material. This is one case in which the two possibilities ate statistically not distinguishable, and the values for the mass of the secondary given in the Table represent the extremes of a range rather than two distinct alternatives.

G107-69 was discussed by Harrington et al (1981) from 15 years of USNO data from two series, the second series being directed towards the binary white-dwarf companion, G107-70. The perturbation was first suspected from a very long periodicity in the parallax itself, which suggested a perturbation with a period close to a year beating against the annual parallax period. We have no way to distinguish between the two possibilities for the mass of the secondary, and this result should probably be regarded as a bit tentative anyway.

With the increased time base for the traditional refractor parallax series, and the increased accuracy in the more modern reflector series, it is possible to detect, from their astrometric perturbations, unseen companions of ever smaller masses. This should yield many more brown dwarfs (if they really exist in large numbers) as time builds up astrometric data bases. This should begin to yield significant data not only on the frequency of brown dwarfs (at least in multiple systems), but also their mass function and its upper and lower limits.

REFERENCES
Christy, J. W, (1978). The Perturbation of G139-29. Astron. J. 83, 1225.
Dahn, C. C., Harrington, R. S., Riepe, B. Y., Christy, J. W., Guetter,
        H. H., Kallarakal, V. V., Miranion, M., Walker, R. L.,
        Vrba, F. J., Hewitt, A. V., Durham, W. S., and Ables, H. D.
        (1982). U. S. Naval Observatory Parallaxes of Faint Stars.
        List VI. Astron. J. 87, 419.
Harrington, R. S. (1985). The Companion to VB8. B. A. A. S. 17, 624.
Harrington, R. S., Christy, J. W., and Strand, K. Aa. (1981). The
        Nearby Quadruple Star G107-69/70. Astron. J. 86, 909.

Harrington, R. S., Kallarakal, V. V., and Dahn, C. C. (1983). Astrometry
         of the Low-Luminosity Stars VB8 and VB10. Astron. J. __88__,
         1038.
McCarthy, Jr., D. W., Probst, R. G., and Low, F. J. (1985). Infrared
         Dtection of a Close Cool Companion to van Biesbroeck 8.
         Ap. J. __290__, L9.
Lippincott, S. L. (1977). Astrometric Analysis of the Unseen Companion
         to Ci18,2354 and Wolf 1062 from Plates Taken with the 61-cm
         Sproul Refractor. Astron. J. __82__, 925.
Lippincott, S. L. (1979). Low-mass Unseen Companions to two Nearby Red
         Dwrafs, CC1228 and Wolf 922. P. A. S. P. __91__, 784.
Strand, K. Aa. (1977). Triple System Stein 2051 (G175-34). Astron. J.
         __82__, 745.
van de Kamp, P., and Worth, M. D. (1972). Parallax and Orbital Motion
         of the Unresolved Astrometric Binary BD+43°4305. Astron. J.
         __77__, 762.

DISCUSSION

Bahcall: How big are the errors on the masses, m_2, in your table?

Harrington: The formal mean errors are very small (15% or less), but
we are at the limits of performance capability of the telescopes in
many cases, so systematic errors are the big worry.

Lunine: Is a mass ratio m_1/m_2 for VB8 of 0.08/0.05 acceptable? How about
0.08/0.07?

Harrington: Yes, although 0.08/0.07 is an extreme value requiring a
rather large orbital inclination.

Probst: 1) Can one get a useful lower limit on the age of the companion
to G107-69 from the cooling age of the white dwarf companion G107-70?
2) Scargle has pointed out a danger in analyzing residuals from a
parallax-proper motion solution for perturbations, viz. a cyclical
perturbation can be partially absorbed into the linear proper motion
fit even when more than one period is covered. Would you comment on the
seriousness of this systematic error.

Harrington: I can not comment on the first point. The second one is
certainly correct and is the reason why well more than one period should
be covered. It should be pointed out that masses are reasonably well
known once acceleration, i.e. curvature, can be measured, and hence
these are usually much better known than either period or amplitude
separately.

Henry: You've shown McCarthy's "negatives"; what about your negatives?
How many stars have been examined astrometrically to find the cases
you are showing? What is the actual abundance of these beasts?

Harrington: We have determined over 800 parallaxes at USNO and have found only 12 unseen companions, of which 4 are on my list here. Of course, most of our stars are farther away and hence these small objects would yield much smaller perturbations.

Roman: Can spectroscopy in the far red rule out the comparable mass cases?

Harrington: Probably, but remember these objects are faint and have very long periods, spectroscopically, making such measurements difficult.

Marcy: Given the 1 arc second separation for VB8B by McCarthy and the theoretical estimates for the mass of the secondary of 0.06, why don't you see astrometric perturbations of order of $\frac{1}{2}$ arc second after your 12 years of observation?

Harrington: Our data is entirely consistent with periods of the order of 50-60 years. Perturbations initially build up like the cosine function, and hence we would expect a rather small amplitude even over this period.

D'Antona: How do you get the primary mass value for VB8A? You can't extrapolate the emperical mass-luminosity relation to 0.08 Suns.

Harrington: You are quite correct. The mass given for VB8A is really only a leap-of-faith value for illustrative purposes, and the mass of the companion scales with the assumed mass of the primary.

Post Script: It was subsequently learned that McAllister had resolved +68°946 in the spring of 1984 with his speckle camera: hence, the equal-mass interpretation applies, and the companion can not be a brown dwarf.

THE SEARCH FOR SUBSTELLAR COMPANIONS TO NEARBY STARS: INFRARED IMAGING FROM THE GROUND AND FROM SPACE

Donald W. McCarthy, Jr.
Steward Observatory, Univ. of Arizona, Tucson, AZ 85721 USA

Abstract. Near-infrared (1.6 to 3.4 microns) speckle interferometry is being used for diffraction-limited imaging of a 10x10 arcsec^2 area around nearby stars in a search for low mass companions. This work provides direct information on the photometric properties and masses of objects near the end of the hydrogen-burning main sequence. Thusfar, 50 stars have been examined with resolution as high as 0.1 arcsec. At least 21 companions have been detected, including the brown dwarf candidate VB 8B. Techniques for image reconstruction resolve the usual ambiguity in position angle and have revealed circumstellar features around one nearby red dwarf. We emphasize the importance of future improvements in infrared detectors and also demonstrate how near-infrared imaging from the Hubble Space Telescope can provide enormous gains in sensitivity.

INTRODUCTION

Until recently, direct detection of low mass companions to nearby stars was technically impractical due to the expected sub-arcsecond separations and the proximity of a more luminous primary star. Today, this perception is changing with the advent of infrared techniques for diffraction-limited imaging from large ground-based telescopes. Now, in relatively short observations, brown dwarf companions can be detected via their own thermal emission. This capability will expand dramatically in the next decade as improvements in detector technology and imaging algorithms are achieved.

Since 1982, I have used "infrared speckle interferometry" to verify low mass objects indicated previously by astrometry and to search for brown dwarf companions to the lowest luminosity red dwarfs (McCarthy 1983, 1984a; McCarthy et al. 1985). This approach has several advantages. First, companions are observed near their black-body peaks. The resulting magnitude difference in the system is 3-4 mag less than at visible wavelengths. Second, atmospheric seeing is significantly improved in the infrared. Third, direct imaging can detect non-astrometric companions such as those in systems with nearly coincident bary- and photo- centers and with revolution periods >50 years. Fourth, speckle and astrometric data yield masses and absolute photometry for the component objects (Lippincott et al. 1983; McCarthy 1984b). In addition to these scientific goals, the study of binary objects provides useful tests of techniques for image reconstruction.

This article summarizes the progress of these searches and provides an update on observations of the brown dwarf candidate Van Biesbroeck 8B (VB 8B). It also addresses the strengths and weaknesses of the present infrared technique and discusses the enormous potential for improvement.

THE INFRARED SPECKLE TECHNIQUE

Speckle interferometry refers to the fact that the telescopic image of an unresolved star is really an interference pattern exhibiting numerous ($\sim$1000) small speckles. The wavefront distortions leading to the speckles are produced by atmospheric turbulence which is highly time-variable such that speckles can only be seen in short (0.03 sec) "exposures". Remarkably, speckles can be visualized as near diffraction-limited images. The name of the game is to preserve the high spatial frequency information inherent in the individual speckles while integrating the light from faint objects.

Figure 1 illustrates the basic ingredients of the infrared speckle technique. Until infrared array detectors are available, we are forced to scan the telescopic image rapidly across a narrow slit in front of a single detector. The scanning motion is supplied by linear tilts of the Cassegrain secondary mirror (f/30,45) and is sufficiently rapid (50-100 arcsec sec^{-1}) to "freeze" the high frequency components of atmospheric seeing. The slit integrates the starlight in one-dimension and multiplexes spatial frequencies in the detector output. The slit has an angular width of $\leq \lambda/D$, where λ is wavelength and D is telescope diameter. In this scheme, the scan motion must be highly linear (distortions $\leq$0.02 arcsec), and the detector must have a flat frequency response over a broad bandwidth (0-500 Hz).

After amplification, the detector signal passes through an anti-aliasing filter, a 12 bit A-D converter, and then into a microcomputer at a rate of 2 kbytes sec^{-1} corresponding to a scan frequency of about 8 Hz. The computer records raw data on magnetic tape, displays scans in real-time, and performs on-line processing for each scan including the following:
1.   coaddition of individual power spectra;
2.   coaddition of phase information (Knox & Thompson 1974);
3.   coaddition of shift-and-add scans (Bates & Cady 1980);
4.   calculation of photometric brightness;
5.   calculation of scan "sharpness".

The above data sets are obtained for both an unresolved point source and the object of interest. In practice, the telescope is moved frequently (every 2-3 minutes) between these two objects in order to reduce effects of slow changes in the Fried parameter, r_o. The point source data provide calibration of the instrumental-atmospheric transfer function.

Figure 1. Schematic diagram of the one-dimensional scanning method presently used in infrared speckle interferometry.

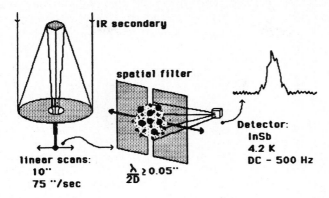

Figure 2. Block diagram of analysis procedures used to achieve diffraction-limited images.

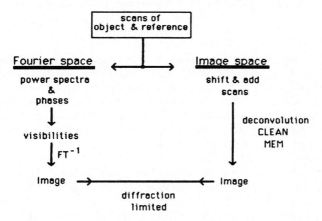

Figure 2 illustrates how this data is used to achieve diffraction-limited imaging. The calibrated power and phase spectra provide the complex visibility function which can be transformed into an image. An image also results from deconvolution of the shift-and-add scans using algorithms such as CLEAN (Högbom 1974), Maximum Entropy Method (Gull and Daniell 1978), and other nonlinear techniques (e.g., Lucy 1974).

Figure 3 illustrates recent results on the nearby star L 789-6 (Gl 866) obtained with the Kitt Peak 3.8-m Mayall telescope at 2.2 microns (McCarthy et al. 1986). The displayed visibility amplitudes and phases indicate a binary source with an angular separation of 0.4 arcsec and a magnitude difference of 0.6 mag. The damping effect in these data is caused by imperfections (peak amplitude 0.06 arcsec) in the scan motion and by atmospheric changes during the scan.

Figure 3. Visibility amplitude and phase versus spatial
frequency for the binary object L 789-6. The damping effect
at increasing frequency results primarily from imperfections
in the scanning motion as described in the text.

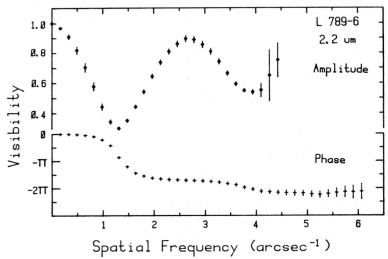

To date, it has been possible to resolve binaries with separations ≥ 0.1
arcsec and magnitude differences ≤ 4.5 mag. The minimum separation is
determined by achievable signal-to-noise at high spatial frequencies.
For bright objects, this limit is imposed by the finite telescope
baseline. The maximum detectable magnitude difference is determined by
the absolute accuracy of visibility measurements and is currently
limited by effects of nonstationary atmospheric seeing. Improved
methods of atmospheric calibration are needed to increase the
measurable magnitude difference and thus to detect less luminous
companions. We are exploring the possibility of measuring scan
sharpness (c.f., Muller & Buffington 1974) to weight individual scans or
to bin them according to instantaneous seeing for subsequent analysis
(Christou et al. 1985). Long period fluctuations occur frequently and
may be associated with the outer scale of turbulence (Mariotti 1983).

ASTROMETRIC BINARIES
        Table 1 summarizes measurements of 19 astrometric binaries
confirmed by infrared speckle interferometry. Each system is listed
according to distance along with its photocentric semi-major axis and
the 2.2 micron absolute magnitude of the companion. An asterisk
signifies that revolution motion has been detected in the infrared.
Most of the astrometric work has been summarized by Lippincott (1978).
Some objects are the result of collaboration such as G250-29 (Dr. E.
Borgmann), LHS 1047 (Dr. P. Ianna), and BD +41°328 (Lippincott et al.
1983).

Table 1.
Confirmed Astrometric Binaries

Object	Distance	Photocentric α	Companion $M_{2.2 \, \mu m}$
Kuiper 84	34.4 pc	0.05 arcsec	~4.7 mag
ζ Cnc C	25.0	0.19	3.3
ζ Aqr B	23.2	0.07	4.4
α Oph	16.7	0.07	4.1
G 96-45	15.9	0.04	~8.4
BD +67°552	14.3	0.16	5.8
BD +41°328	12.8	0.13	6.6
CC 1299	12.3	0.44	7.9
G 250-29	11.6	0.06	7.3
Ross 52	10.4	~0.10	5.9
Wolf 1062*	10.3	0.03	8.4
χ¹ Ori	10.1	0.10	6.1
Wolf 922*	8.2	0.04	8.7
μ Cas	7.7	0.19	7.4
CC 986*	7.4	0.05	9.6
VB 8*	6.5	---	12.8
LHS 1047*	5.5	0.07	9.4
G 208-44	4.8	~0.20	10.2
Ross 614*	4.0	0.31	9.2

* Revolution motion detected.

The table brings out two items of interest. First, the confirmed
binaries all have photocentric semi-major axes >0.03 arcsec. Second,
the intrinsically faintest companions are detected only within 10
parsecs. This effect reflects the fact that the faintest companions can
best be detected around the lowest luminosity, hence nearest, red
dwarfs. In addition, it is interesting to note that all the infrared
companions detected to date (Tables 1,4) have angular separations <1
arcsec.

In principle, the speckle and astrometric measurements of a binary
system can be combined to derive the individual masses through Kepler's
Third Law. The angular separation and parallax yield the linear
separation which is used to scale the photocentric orbit. The infrared
brightness difference and color provide an estimate of the fractional
luminosity in the visible. In practice, errors in these quantities are
magnified in the result, leading to large uncertainties in the masses
(Lippincott et al. 1983; McCarthy 1984b). A summary of all the infrared
measurements including mass determinations is in preparation.

Table 2.
Unconfirmed Astrometric Binaries

Object	Distance	Photocentric α	Companion Upper Limit $M_{2.2\,\mu m}$
G 146-72	33.3 pc	0.03 arcsec	9.8 mag
γ Gem	32.3	0.03	
36 UMa	12.7	0.03	6.9
G 139-29	12.5	0.06	12.5
CC 1228	12.2	0.01	10.0
G 107-69	11.2	0.01	12.0
BD +66°34A	10.0	0.13	10.2
G 24-16	8.7	0.03	11.5
BD +6°398	7.8	0.22	7.8
VB 10	5.8	0.02	14.0
Stein 2051A	5.5	0.07	11.0
BD +43°4305	5.0	0.02	10.7
BD +68°946	4.7	0.04	10.2
Barnard's $*$	1.8	0.01	12.3

Table 2 lists astrometric binaries thusfar unconfirmed in the infrared. Existing measurements provide upper limits on the absolute brightnesses of any companions. In a few cases, these limits preclude companions like VB 8B. Most of the unconfirmed objects have photocentric semi-major axes $\leq$0.03 arcsec. Although several have suspected substellar companions which should be undetectable in the infrared, it is also possible that these weak perturbations are not real. The stronger perturbations for BD +6°398, BD +66°34A, and G 139-29 indicate that companions should be detectable, and we are continuing to study these objects. VB 10 and Stein 2051A are not strong astrometric candidates.

Table 3 summarizes the lowest luminosity objects known at 0.55 and 2.2 microns. Asterisks indicate objects discovered as close companions in the infrared. Absolute visual magnitudes are taken from Dahn et al. (1985) or estimated (in parentheses) from the 2.2 micron magnitudes using empirical relationships found by Veeder (1974). Clearly, it is important to allow for binary companions in any determination of the luminosity function.

SURVEY OF NEARBY STARS

In 1983, I began an infrared speckle survey of nearby stars in order to detect substellar companions and any companions possibly missed by astrometry. Because of the current ≤ 4.5 mag limit on detectable brightness difference, the primary targets in the search for brown dwarfs are the lowest luminosity red dwarfs. Consequently, Van Biesbroeck 8 and 10 were two of the first objects studied.

Table 4 lists the results to date. Two objects have been resolved for the first time. The L 789-6 (G1 866) system has a magnitude difference of only 0.6 mag at 2.2 microns. The companion may have been missed by parallax studies because the barycenter and photocenter are nearly coincident. Ten nearby stars are unresolved. The resulting upper limits on the absolute brightnesses of possible companions generally preclude objects resembling VB 8B ($M_K=12.8$). Extending these limits requires more accurate visibility measurements and, therefore, improved calibration of atmospheric seeing.

Table 3.
Lowest Luminosity Objects

| Object | Absolute Magnitudes | |
	$M_{2.2 \mu m}$	M_V
RG 0050-2722	10.5	!9.5
LHS 2924	10.5	19.6
* CC 986B	9.6	(16.6)
VB 8A	9.8	17.7
* VB 8B	12.8	(30:)
VB 10	10.0	19.0
* LHS 1047B	9.4	(16.2)
G 9-38	(9.4)	16.3
G 208-44A	9.2	15.1
* G 208-44B	10.2	(17.6)
G 158-27	9.1	15.4
* L 1159-16B	9.0	(15.5)
LHS 292	(10.0)	17.3
* Ross 614B	9.2	16.6
G 51-15	9.5	17.0
* L 789-6B	9.0	(15.6)
Wolf 359	9.3	16.6

* Infrared companion.

Table 4.
Additional Nearby Stars

Object	Distance	Companion $M_{2.2 \mu m}$
Resolved:		
L 1159-16	4.5 pc	11.6 mag
L 789-6	3.3	9.0
Unresolved:		
70 Oph	4.9	≥ 6.5
G 208-45	4.8	≥ 12.9
Wolf 424	4.3	≥ 12.6
BD +5°1668	3.8	≥ 11.0
G 51-15	3.6	≥ 13.5
Σ 2398A	3.5	≥ 11.1
Ross 128	3.3	≥ 12.0
Ross 248	3.1	≥ 12.5
L 726-8	2.6	≥ 12.2
Wolf 359	2.3	≥ 13.3

UPDATE ON VAN BIESBROECK 8B
        I reobserved VB 8B in May and June 1985, in order to confirm
and expand upon the original observations. Results at 1.6 and 2.2
microns still show an object with the expected brightness and color.
The angular separation is still about 1 arcsec; however, the position
angle has increased to about 45 (225) degrees. This motion does not
mimic the proper motion of VB 8, so the possibility of a chance
background object is eliminated.

Problems with weather and detector performance precluded observations at
3.4 microns where VB 8B should be brighter relative to the primary star.
Observations at this wavelength are difficult because of increasing
thermal background. Below 1.6 microns, VB 8B should be much fainter and
essentially invisible given the 4.5 mag limit in detectable brightness
ratio. Detector sensitivity currently restricts measurements to the
standard photometric bands.

CIRCUMSTELLAR ENVELOPES
        A byproduct of the search for low mass companions has been
the detection of circumstellar emission from the red dwarf Lalande 21185
(Gl 411). Figure 4 shows one-dimensional images of this object and an
unresolved star, HR 4280. Lalande 21185 exhibits an asymmetrical
envelope containing about 10 percent of the total light emitted at 2.2
microns. The Steward 2.3-m telescope achieves a linear resolution of
0.65 AU at the distance of this object (2.5 pc).

        Figure 4. Diffraction-limited images of Lalande 21185 and
        HR 4280 obtained from inversion of the visibility function.

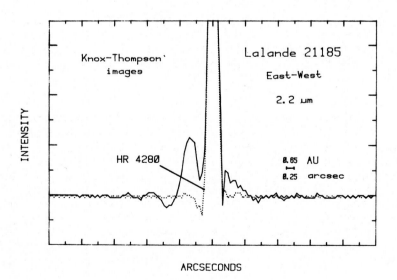

The nature of this envelope remains unclear. It is unlikely to be caused by companion objects since astrometry (Hershey & Lippincott 1982) allows a companion of at most two Jupiter masses. IRAS results do not reveal a substantial infrared excess above the stellar continuum. Possibly, the envelope is a remnant of previous stellar flaring or represents scattered light from a localized disk of dust. In fact, the speckle work shows some evidence for variability. If the scattering interpretation is correct, then flare activity in such objects might produce "noise" in astrometric data as the photocenter changes slightly with time.

EXISTING LIMITATIONS AND FUTURE IMPROVEMENTS
The present infrared speckle technique requires that a scan through the entire image be completed within the speckle lifetime and that the scan motion and telescope tracking be highly linear. Violation of these conditions causes attenuation and modification of the transfer function as well as production of energy beyong the spatial frequency cutoff of the telescope (Aime et al. 1980; Leinert & Dyck 1983). As illustrated in Figure 3, these conditions are not precisely satisfied in our present equipment resulting in degradation of the visibility data.

In addition, the scanning method suffers other inefficiencies including:
1. short dwell time on the complete telescopic image;
2. additional signal losses caused by:
    a. diffraction losses at slit
    b. attenuation due to slit transfer function
    c. blurring of speckles along slit
3. large noise bandwidths;
4. one-dimensional information only.

These problems can be remedied by two-dimensional array detectors which stare at the complete image without scanning and which can be read out at high speeds to provide short exposures. We are seeking to make this transition, but it is expensive and will generate a large amount of data in a short time.

NEAR-INFRARED IMAGING FROM THE HUBBLE SPACE TELESCOPE
The present investigation demonstrates the advantages of searching for substellar companions in the near-infrared. However, this work has also demonstrated the calibration difficulties imposed by variable atmospheric seeing. The best sensitivity for such observations can be obtained from above the atmosphere where a telescope concentrates star light into a single speckle. Combined with excellent optics and telescope tracking, such observations could utilize long integrations and the full dynamic range to detect much larger brightness differences. It would be possible to image lower mass brown dwarfs around brighter stars in a fraction of the required ground-based observing time.

The Hubble Space Telescope (HST) provides a unique opportunity to capitalize on these advantages in the near future. Although HST is presently equipped with visible and ultraviolet instrumentation, NASA has announced an opportunity for second-generation instrumentation to be installed as early as 1992. A near-infrared camera and spectrometer has been proposed to take full advantage of the near-perfect optics and tracking capabilities of HST.

Coupled to modern infrared array detectors, this instrument could image brown dwarf companions around solar-type stars in seconds. Calculations show that VB 8B could be detected with improved resolution in a fraction of a second compared with two hours of ground-based integration. This capability can be optimized with coronographic optics to block out direct and scattered light from the central star.

ACKNOWLEDGMENTS
The author gratefully acknowledges assistance from Drs. R. Probst, R. Joyce, F. Low, G. Rieke, M. Rieke, and M. Cobb. This project receives financial support from the National Science Foundation under grant AST-8218782.

REFERENCES
Aime, C., Kadiri, S., and Ricort, G. (1980). The Influence of Scanning Rate in Sequential Analysis of a Speckle Pattern. Application to Speckle Boiling. Opt. Comm., 35, 169.

Christou, J. C., Cheng, A. Y. S., Hege, E. K., and Roddier, C. (1985). Seeing Calibration of Optical Astronomical Speckle Interferometric Data. A. J., 90, 2644.

Dahn, C. C., Liebert, J., and Harrington, R. S. (1985). LHS 292 and the Luminosity Function of the Nearby M Dwarfs. preprint.

Gull, S. F., and Daniell, G. J. (1978). Image Reconstruction from Incomplete and Noisy Data. Nature, 272, 686.

Hershey, J. L., and Lippincott, S. L. (1982). A Study of the Intensive 40-yr Sproul Plate Series on Lalande 21185 and BD +5°1668. A. J., 87, 840.

Högbom, J. A. (1974). Aperture Synthesis with a Non-Regular Distribution of Interferometer Baselines. Astron. Astrophys. Suppl., 15, 417.

Leinert, C., and Dyck, H. M. (1983). Speckle Interferometry Degraded by Irregular Motion of a Scanning Telescope Mirror. Appl. Opt., 22, 2403.

Lippincott, S. L. (1978). Astrometric Search for Unseen Stellar and Sub-Stellar Companions to Nearby Stars and the Possibility of Their Detection. Sp. Sci. Rev., 22, 153.

Lippincott, S. L., Braun, D., and McCarthy, D. W. (1983). Astrometric and Infrared Speckle Analysis of the Visually Unresolved Binary BD +41°328. Pub. A. S. P., 95, 275.

Lucy, L. B. (1974). An Iterative Technique for the Rectification of
    Observed Distributions. A. J., 79, 745.
Mariotti, J. M. (1983). Experimental Results on Atmospheric Turbulence
    Obtained with an Infrared Speckle-Interferometer. Opt.
    Acta, 30, 831.
McCarthy, D. W. (1983). Near-Infrared Imaging of Unseen Companions to
    Nearby Stars. In The Nearby Stars the Stellar Luminosity
    Function, ed. A. G. D. Phillips and A. R. Upgren, pp. 107-
    112.
McCarthy, D. W. (1984a). Infrared Speckle Interferometry: A Sensitive
    Technique for Physical Measurements of Unseen Companions to
    Nearby Stars. In Astrometric Techniques, ed. H. K.
    Eichhorn, in press.
McCarthy, D. W. (1984b). Mass Measurement of the Components of Mu Cas.
    A. J., 8, 433.
McCarthy, D. W., Probst, R. G., and Low, F. J. (1985). Infrared
    Detection of a Close Cool Companion to Van Biesbroeck 8.
    Ap. J. (Letters), 290, L9.
McCarthy, D. W., Probst, R. G., and Cobb, M. L. (1986). in preparation.
Muller, R. A., and Buffington, A. (1974). Real-Time Correction of
    Atmospherically Degraded Telescope Images Through
    Sharpening. J. O. S. A., 13, 1200.
Veeder, G. J. (1974). Luminosities and Temperatures of M Dwarf Stars
    From Infrared Photometry. A. J., 79, 1056.

DISCUSSION

Lunine: What are the prospects for detecting VB 8B at L band and beyond, _i.e._, getting three or four color data?

McCarthy: As I mentioned earlier, observations at 3.4 microns are feasible, and we are attempting them. However, they are difficult because of the influence of thermal background radiation. Shorter wavelengths are more difficult because of the steep Planck function for VB 8B and because of the difficulty in atmospheric calibration for brightness ratios >4.5 mag.

Bahcall: If the proposed instrument for the Hubble Space Telescope were utilized on the ground, what benefits would be obtained?

McCarthy: The array detectors and sensitive read-out electronics would improve photometric sensitivity. However, there would be no improvement in our ability to calibrate atmospheric seeing. Thus, accurary in measurements of absolute visibilities would still be poor, and the 4.5 mag restriction on measurable brightness ratio in binary systems would remain.

Henry: In doing your Fourier analysis, you do East-West cuts and North-South cuts. Would you significantly reduce errors or improve results if you did additional cuts at other angles?

McCarthy: Other position angles would provide improved sampling of the (u,v) plane and would lead to better image reconstruction. In binary systems, we could measure more accurately the separation vector. Unfortunately, at present, each position angle requires a separate observation. Array detectors would be advantageous.

Harrington: The tabulated low luminosity objects are, except for VB 8B, more probably very red dwarfs, not brown dwarfs.

McCarthy: Given the likely mass of VB 8B presented at this conference, you are probably correct. However, the status of LHS 2924, in particular, is still unclear.

Zinnecker: Can you say a little more about the instrument with which you obtain the two-dimensional 2.2 micron image of Lalande 21185?

McCarthy: These images were preliminary exposures taken by Dr. M. Rieke with a 32x32 HgCdTe array camera at the Steward Observatory 2.3-m telescope. The exposure tiimes were about 0.1 sec so the speckle structure is mostly smoothed out.

Probst: Do you expect problems due to confusion from background
   sources--stars and galaxies--using the Space Telescope at two
   microns?

McCarthy: Source confusion is likely to a problem, just as it was
   potentially for VB 8B itself. However, the substantial proper
   motion of these nearby stars should help eliminate this problem
   entirely.

Tarter: In contemplated use of SIRTF, do you need to rely on
   "super-resolution" techniques and, if so, is the signal-to-noise
   from these types of systems sufficient?

McCarthy: SIRTF will achieve very high signal-to-noise observations
   suitable for super-resolution analysis. Even so, SIRTF's ability to
   resolve brown dwarf companions will be limited by the diameter and
   quality of the primary mirror and by tracking stability. My own
   feeling is that SIRTF can only look for companions around the very
   nearest stars and that, instead, it should utilize its sensitivity
   to conduct a sensitive sky survey for free-floating brown dwarfs.

CONSTRAINTS ON THE BROWN DWARF MASS FUNCTION FROM OPTICAL AND INFRARED SEARCHES

R. G. Probst
National Optical Astronomy Observatories, P.O. Box 26732,
Tucson, AZ 85716 USA

Abstract. Photometric surveys of faint proper motion stars
and searches for infrared binary companions have identified
a few very low luminosity objects. I consider how these
searches may constrain the brown dwarf mass function. An
astrophysically plausible brown dwarf population is defined
which yields a dark mass density = 0.5 x the observed density. Using the sensitivity and other limits of various
surveys, I obtain the expected numbers of observable brown
dwarfs from the model population for comparison with actual
results. The deepest available proper motion survey may
contain a few brown dwarfs, but a deeper survey would be
needed to yield a statistically meaningful result. Binary
searches are more promising. Null detections previously
reported can be attributed to insufficient areal coverage.
Reasonable improvement in search protocol could yield
statistically significant tests of the brown dwarf mass
function.

INTRODUCTION

The nature and amount of unobserved matter in the Galactic
disk is a longstanding problem with an equally old favorite solution:
a substantial population of very low mass stars and substellar brown
dwarfs (BDs). During the past few years I have conducted an optical
and infrared photometric survey of previously identified faint red
proper motion stars (Probst 1985), one goal of which was to find new
very low luminosity objects as BD candidates. McCarthy (1986) and
others (Jameson et al. 1983; Skrutskie et al. 1986) have searched for
very low luminosity binary companions to known stars with infrared
detectors. Both methods have had some degree of success (Probst and
Liebert 1983; McCarthy et al. 1985). I wish to consider whether these
complementary strategies for field objects and binary companions can
constrain the characteristics of BDs as a population in a useful way.

My approach is intended to be illustrative, not exhaustive. I construct
a single, astrophysically plausible BD mass function which yields a
significant amount of dark mass. Analytical relations are used to
obtain the observed BD flux as a function of effective wavelength.
Using the sensitivity, spatial coverage, and other limits of various

surveys, I derive the expected numbers of observable BDs from the model population for comparison with the actual results. Suggestions are made as to how these searches might be improved to increase the expected numbers of detections and so improve the statistical significance.

THE MODEL BD POPULATION
To determine a plausible mass function and space density, I convert the Wielen et al. (1983) luminosity function LF to a mass function MF using the mass-luminosity slope from Miller and Scalo (1979), then fit a power law over the range $0.2 \lesssim M/M_\odot \lesssim 1.0$. The resulting fit is shown in Figure 1. The mass range is chosen such that the source LF is integrated over the lifetime of the Galactic disk, and neither the LF nor the mass-luminosity conversion are in serious doubt. The model MF is terminated at $0.01\ M_\odot$, a value suggested by recent work on fragmentation and star formation (Boss 1986).

The BD formation rate is assumed constant over 1×10^{10} yr (Mould 1982; Twarog 1980). To determine the BD flux as a function of mass M and age t, I use analytical relations for radius R and effective temperature T_e given by Stevenson (1986). Resulting values for luminosity L and T_e as functions of (M, t) compare reasonably well with recent numerical calculations (D'Antona and Mazzitelli 1985; Nelson et al. 1985). The BD (M, t) plane is represented as a grid in log M and t, with the space density determined for each bin from the model mass function.

This model BD population is _not_ defined by any a priori constraint on the mass density. Integrating over $0.2 \gtrsim M/M_\odot \gtrsim 0.01$ and subtracting the

Figure 1. The model fit to the mass function of nearby stars.

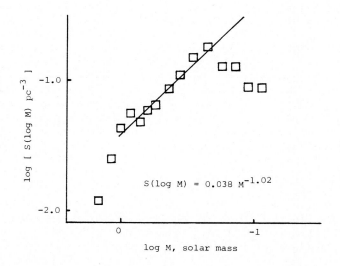

$$S(\log M) = 0.038\ M^{-1.02}$$

mass density of known stars gives

$$\frac{\text{mass density of dark matter}}{\text{mass density of luminous matter}} = 0.5$$

with the bulk of the dark matter density due to BDs. The value of 0.5 is within the range of 0.5 to 1.5 determined by Bahcall (1986). The mean BD mass is 0.024 $M_\odot$ and the total space density is 1.6 BD pc^{-3}. This mass function predicts a substantial number of undiscovered stars with $0.08 \lesssim M/M_\odot \lesssim 0.2$. Within a 5 pc horizon, the number of undiscovered stars $\approx$ number of known stars ≈ 35. This seems uncomfortably high. However, new stars are still being found within 5 pc, both as field stars (Harrington et al. 1983) and as binary companions (McCarthy 1986). Also Krisciunas (1977) argues from velocity dispersions that for a 10 pc horizon, the total number of stars ≈ 2.5 x the known number.

SEARCHES AMONG PROPER MOTION STARS

I first test this model BD population against optically visible proper motion stars. The most comprehensive motion survey available is that of Luyten (1963 et seq.) conducted with Palomar Schmidt plates scanned by a specially constructed machine. It covers ~70% of the sky to a red magnitude limit $R_{lim} \sim 19\text{-}20$ and motion limits $0.1 \lesssim \mu \lesssim 2.5$ arcsec yr^{-1}. Photometric and astrometric followup work has been limited to the ~3600 objects with $\mu \gtrsim 0.5$ arcsec yr^{-1} for which charts are available (Probst 1985; Dahn et al. 1986).

The completeness of the Luyten survey as a function of apparent magnitude is not well established. For the present I take it to be complete to R = 19.0 (Kron system) for the determination of flux-limited search radii r(M, t). In computing R band fluxes for BDs I allow a flux depression $\Delta R = 2.9$ magnitudes below the blackbody line, as observed in the coolest known stars (Probst and Liebert 1983, Figure 4). Since detectable BDs are not found to be much cooler, this is a reasonable approximation.

Completeness as a function of proper motion is likewise unclear, but appears very high for $\mu > 0.2$ arcsec yr^{-1} (Luyten 1975). I assume it to be 100% for $\mu \gtrsim 0.5$ arcsec yr^{-1}. The motion constraints set outer and inner limits $r(\mu)$ on the search radii. Since the average tangential velocity $\bar{V}(t)$ increases with stellar age, $r(\mu)$ will vary with time across the BD (M, t) plane. From Wielen (1977) I take

$$\bar{V}(t) = [\, 50 + 2.92\text{x}10^7\, t_{yr}\,]^{0.5} \text{ km s}^{-1} \quad .$$

The search volume and corresponding number of detections is then obtained for every (M, t) bin.

Summing over all (M, t) gives an expected number of detections $\langle N \rangle = 6 \pm 2$ (Poisson uncertainty). As a few of these may be obscured in binaries, the number of isolated objects expected in the motion survey is 3 or 4. Mean values of observable quantities are $\bar{\mu} = 0.7$, $\bar{R} = 18.0$, and $\bar{T}_e = 2000$ K. While it is suggestive that the observed number of cool objects found in an incomplete search of the motion sample is of the same order of magnitude, clearly these numbers are too small to be statistically useful.

How might the proper motion survey be improved to yield a larger number of possible detections, particularly for older, cooler BDs? The outer search radii are set by the flux limit for all but the youngest and hottest BDs, so decreasing the lower limit on motion has little effect while a fainter limiting magnitude greatly increases the search volume. Increasing the upper motion limit decreases the inner search radii and thus increases the search volume for older BDs of higher tangential velocity, although the effect is not as important as a fainter flux limit.

The Palomar Sky Survey is now being repeated to significantly fainter limiting magnitudes (Schombert et al. 1985). Suppose this is used to perform a proper motion survey with limiting R = 21.0 and motion $0.5 \le \mu \le 5.0$ arcsec yr^{-1} (requiring ~ 5 yr between plate pairs). The expected number of BDs from my model population is then $\langle N \rangle = 22 \pm 5$, with mean $\bar{\mu} = 0.7$, $\bar{R} = 20.0$, and $\bar{T}_e = 1800$ K. This number is large enough to be interesting. Even allowing for binaries, a null detection of BDs would reject my trial mass function at the 3σ level and would be an even more stringent test of steeper mass functions. Contrariwise, any BDs found would be sufficiently bright and spatially isolated for observation at higher spectral resolution, providing useful data for theorists (cf. Lunine 1986; Stevenson 1986). The requisite motion survey and followup work would unquestionably be a major undertaking. The availability of high-speed automatic measuring machines and sensitive array detectors on efficient moderate-aperture telescopes make the notion tractable.

INFRARED BINARY COMPANION SEARCHES

BD temperatures are expected to be < 2000 K over most of the (M, t) plane, so infrared detection at $\lambda > 1$ micron is better matched to the spectral energy distribution than are deep optical surveys. It is presently infeasible to survey large areas with single detectors or small arrays, so infrared observers have searched for BDs as binary companions to known nearby stars. Jameson et al. (1983) and Skrutskie et al. (1986) report null results, while McCarthy et al. (1985) have detected the only firm BD found to date. I now test my model BD population against this search strategy.

For a target star of known distance but unknown age and a search to a given limiting magnitude, some portion of the BD (M, t) plane is observable. I take a 2.2 micron limiting magnitude K = 15.0, corresponding to the reported sensitivities of Jameson et al. and Skrutskie et al., and assume BDs are blackbodies at 2 microns (Reid and Gilmore 1984). The target visible stars are assumed uniformly distributed in age over 1×10^{10} yr. The BD companion MF is assumed to have the same shape as the field BD function, as is observed for stars (Abt 1979). The number densities n(log M, t) determined previously are renormalized such that the sum over all M and t bins = 1.00; this is equivalent to assuming one BD companion per target star. The number P of observable companions per star under these assumptions ranges from 0.24 at 3.7 pc to 0.12 at 9.8 pc.

Since binary searches have finite inner and outer angular limits, P must be further multiplied by the probability Q that the BD will lie within the search area. Lacking any direct information for BDs, I estimate Q from the distribution of projected radial separations for 370 visual binaries in the Gliese catalog (Gliese 1969; Gliese and Jahreiss 1979). This distribution is shown in Figure 2. Assuming it applies to secondaries of very low mass may seem a bit desperate, but the distribution does not show any noticeable correlation with absolute magnitude or component magnitude difference over the wide range in mass of source binaries.

Jameson et al. (1983) scanned a 65x65 arcsec field around 21 stars, with an inner limiting radius of 10 or 13 arcsec. I have determined P and Q for each star in their sample. The means are $\overline{P} = 0.21$, $\overline{Q} = 0.24$ and the expected number of detections is $\langle N \rangle = 1 \pm 1$. The Q's are small because of the large central hole in the search area. Skrutskie et al. (1986) have imaged 14x14 arcsec fields around 60 stars within 12 pc, with a small central hole of 2 arcsec radius in the search field. Lacking identifications of the stars at this writing, I assume them to be uniformly distributed within 12 pc and find mean $\overline{P} = 0.15$, $\overline{Q} = 0.26$,

Figure 2. The distribution of projected radial separations in AU for visual binaries in the Gliese catalog.

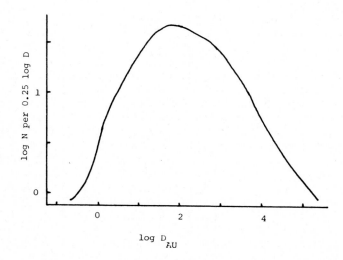

and $\langle N \rangle = 2 \pm 1$. A more centrally condensed distribution gives
$\langle N \rangle = 3 \pm 2$. Here the low Q's are due to the small outer search radius.
McCarthy's (1986) ongoing survey of nearby stars covers 10x10 arcsec
with a flux limit set by the maximum detectable magnitude difference
$\Delta K = 4.5$. As it is incomplete I have not determined $\langle N \rangle$, but expect
it to be similarly small.

The small numbers of detections, predicted and observed, in these
searches provide no useful constraint on the test BD mass function.
To obtain a large value of Q, the outer search radius must be fairly
large and the inner limit as small as possible. For typical distances
~ 7 pc, an outer radius of 30 arcsec and no central hole gives $Q \sim 0.7$.
This radius is a convenient break point, beyond which the return in
additional detections for incremental increases in radius decreases
rapidly. Direct imaging with an array plus interferometric methods
applied to the target star would search such an area thoroughly and
efficiently. A sample of 100 stars uniformly distributed within 10 pc
and searched as before to K = 15.0 would give mean $\bar{P} = 0.16$, $\bar{Q} = 0.71$,
and $\langle N \rangle = 11 \pm 3$. A fainter limiting magnitude and/or search at longer
wavelength would further increase $\langle N \rangle$. This search strategy is very
attractive, as it is reasonable both technically and in terms of
telescope time. Interpretation of results is hampered by the necessity
of assuming a ratio of BD companions per star, which may be determined
by physical conditions independent of the MF itself.

A binary search targeted on known young stars ($t < 1 \times 10^9$ yr) would
greatly increase $\langle N \rangle$, other factors being constant. A step in this
direction has been taken by the Skrutskie et al. observations of several
Pleiades cluster members. But age indicators for field dwarfs are
uncertain, while mass segregation in open clusters (van Leeuwen 1980)
may produce a BD companion MF which is not representative of the field.

CONCLUSIONS

Photometric and other followup observations of faint proper
motion stars are a reasonably efficient means of identifying very low
luminosity stellar objects and perhaps a few young, hot BDs (and also
extreme subdwarfs and cool white dwarfs). But the magnitude limit of
the deepest available motion survey does not provide a significant
test of my trial BD mass function, or of any function with a similar
number density of massive BDs ($M \gtrsim 0.05 \, M_\odot$). Improving this situation
will require a major new initiative.

Infrared binary star searches conducted to date are likewise inconclusive, but the approach is promising. Complete coverage of moderate
angular fields, around a reasonable number of stars, to low but attain-
able flux levels, can provide a significant test for BD mass functions
down to $\sim 0.02 \, M_\odot$. Available instrumentation is adequate to the task,
and the telescope time required is not excessive.

REFERENCES

Abt, H. A. (1979). The Frequencies of Binaries on the Main Sequence. A. J. 84, 1591.

Bahcall, J. N. (1986). Dark Matter in the Galactic Disk. In Dark Matter in the Universe, ed. J. Kormendy, in press.

Boss, A. P. (1986). Theoretical Determination of the Minimum Proto-stellar Mass. In Astrophysics of Brown Dwarfs, ed. M. Kafatos and R. S. Harrington, in press.

Dahn, C. C., Liebert, J., Boeshaar, P. C., Monet, D., and Probst, R. G. (1986). In preparation.

D'Antona, F., and Mazzitelli, I. (1985). Missing Mass in the Solar Neighborhood. Preprint.

Gliese, W. (1969). Catalogue of Nearby Stars. Veroff. Astron. Rechen-Inst. Heidelberg, No. 22.

Gliese, W., and Jahreiss, H. (1979). Nearby Star Data Published 1969-1978. Astron. Ap. Suppl. 38, 423.

Harrington, R. S., Dahn, C. C., and Liebert, J. W. (1983). LHS 292: A New Late-Type Red Dwarf within Five Parsecs of the Sun. In The Nearby Stars and the Stellar Luminosity Function, ed. A. G. D. Philip, p. 339. Schenectady, New York: L. Davis Press.

Jameson, R. F., Sherrington, M. R., and Giles, A. B. (1983). A Failed Search for Black Dwarfs as Companions to Nearby Stars. M. N. R. A. S. 205, 39P.

Krisciunas, K. (1977). Toward the Resolution of the Local Missing Mass Problem. A. J. 82, 195.

Lunine, J. I. (1986). Compositional Indicators in Infrared Spectra of Brown Dwarfs. In Astrophysics of Brown Dwarfs, ed. M. Kafatos and R. S. Harrington, in press.

Luyten, W. J. (1963). Proper Motion Survey with the 48-Inch Schmidt Telescope, I. Minneapolis: University of Minnesota.

_____. (1975). Proper Motion Survey with the 48-Inch Schmidt Telescope, XL. Minneapolis: University of Minnesota.

McCarthy, D. W. (1986). The Search for Substellar Companions to Nearby Stars. In Astrophysics of Brown Dwarfs, ed. M. Kafatos and R. S. Harrington, in press.

McCarthy, D. W., Probst, R. G., and Low, F. J. (1985). Infrared Detection of a Close Cool Companion to Van Biesbroeck 8. Ap. J. (Letters) 290, L9.

Miller, G. E., and Scalo, J. M. (1979). The Initial Mass Function and Stellar Birthrate in the Solar Neighborhood. Ap. J. Suppl. 41, 513.

Mould, J. R. (1982). Stellar Populations in the Galaxy. Ann. Rev. Astron. Ap. 20, 91.

Nelson, L. A., Rappaport, S. A., and Joss, P. C. (1985). On the Nature of the Companion to Van Biesbroeck 8. Nature 316, 42.

Probst, R. G. (1985). A Photometric Survey of the Stars of Largest H. Bull. A. A. S. 17, 553.

Probst, R. G., and Liebert, J. (1983). LHS 2924: A Uniquely Cool Low-Luminosity Star with a Peculiar Energy Distribution. Ap. J. 274, 245.

Reid, N., and Gilmore, G. (1984). Infrared Photometry and the H-R
        Diagram for Very Low Mass Dwarfs. M. N. R. A. S. $\underline{206}$, 19.
Schombert, J., Mould, J., Sargent, W. L. W., Kowal, C., Maury, A., and
        Brucato, R. (1985). The Palomar Sky Survey II. Bull. A.
        A. S. $\underline{16}$, 907.
Skrutskie, M. F., Forrest, W. J., and Shure, M. A. (1986). A Search
        for Low-mass Companions of Stars within 12 Parsecs of the
        Sun. In Astrophysics of Brown Dwarfs, ed. M. Kafatos and
        R. S. Harrington, in press.
Stevenson, D. J. (1986). High Mass Planets and Low Mass Stars. In
        Astrophysics of Brown Dwarfs, ed. M. Kafatos and R. S.
        Harrington, in press.
Twarog, B. A. (1980). The Chemical Evolution of the Solar Neighborhood.
        Ap. J. $\underline{242}$, 242.
van Leeuwen, F. (1980). Mass and Luminosity Function of the Pleiades.
        In Star Clusters, ed. J. E. Hesser, pp. 157-163. Boston:
        Reidel.
Wielen, R. (1977). The Diffusion of Stellar Orbits Derived from the
        Observed Age-Dependence of the Velocity Dispersion. Astron.
        Ap. $\underline{60}$, 263.
Wielen, R., Jahreiss, H., and Kruger, R. (1983). The Determination of
        the Luminosity Function of Nearby Stars. In The Nearby
        Stars and the Stellar Luminosity Function, ed. A. G. D.
        Philip and A. R. Upgren, pp. 163-170. Schenectady, New
        York: L. Davis Press.

DISCUSSION

D'Antona: You derive the main sequence mass function from Wielen's
luminosity function plus a power law mass-luminosity relation. This is
why your mass function declines at low mass.

Probst: The uncertainty in the conversion from luminosity to mass is
why I limited the fit of my model MF to $M \gtrsim 0.2$ solar masses. A rising
MF below this value would reduce the rather large number of undiscovered
stars predicted by the model MF.

PROPERTIES OF STELLAR OBJECTS NEAR THE MAIN SEQUENCE MASS LIMIT

J.W. Liebert
Steward Observatory, Univ. of Arizona

C.C. Dahn
U.S. Naval Observatory, Flagstaff station

1   INTRODUCTION

In collaboration with P.C. Boeshaar and R. Probst, we have been studying a sample of stars of large proper motion, extracted from the LHS Catalogue (Luyten 1979). The sample consists of all stars with proper motions $\geq 0\overset{..}{.}8$ yr^{-1}, at declinations north of -20° (approximately the northern 2/3 of the sky). For reasons which are well known, such a sample is optimum for the selection of (1) nearby stars of low luminosity, and (2) high velocity subdwarf stars in the more extended solar neighborhood. We shall concentrate on the former at this conference, and shall make use of a set of preliminary CCD trigonometric parallaxes obtained by D. Monet and collaborators at the U.S. Naval Observatory. In Section 2, the luminosity function of the lower end of the main sequence derived from this sample is compared with other recent determinations. In Section 3, the colors, kinematics and spectroscopic properties of the lowest luminosity stellar objects are discussed, with an emphasis on identifying methods for determining empirically which objects might be of substellar mass (brown dwarfs). In this we make use of radial velocities obtained in collaboration with M. Giampapa (Giampapa & Liebert 1986).

2   THE LUMINOSITY FUNCTION FROM THE LHS SAMPLE

This sample of large proper motion stars includes nearly 50 stars with $M_V \geq +15$, a large fraction of which were not previously analyzed. The faintest object in the sample is LHS 2924 (Probst & Liebert 1983; Liebert et al. 1984; Monet & Dahn 1984), for which the latest CCD parallax update ($\pi_{abs} = 0\overset{..}{.}092 \pm 0\overset{..}{.}001$ m.e.; Monet 1985) gives $M_V = +19.6 \pm 0.01$.

In view of the extreme selection effects in favor of high velocity stars resulting from this sample, it is necessary to derive and interpret a luminosity function of these objects with utmost caution. A complete discussion will be given in a forthcoming paper (Dahn et al. 1986a). For now, it is appropriate to summarize the results for the lower main sequence of disk stars (spectroscopic dwarfs) and to compare them with other recent determinations. Our analysis makes use of the 1/Vmax method of Schmidt (1975), with the volumes defined by (1) a faint apparent magnitude limit, and (2) both lower and upper proper motion limits. The former is taken as around $m_R \sim 18.0\text{-}18.5$ (Luyten photo-

graphic red Palomar Sky Survey estimate). The proper motion is
restricted to $0\overset{''}{.}80$-$2\overset{''}{.}00$ yr^{-1}. The former is our arbitrary lower limit
defining this subset of the LHS Catalogue; the latter is the practical
upper limit for motions found on CDC machine-measured plates for the
great majority of the Luyten Palomar Survey (Luyten & La Bonte 1973).
Actually, there are many fairly bright stars with motions > $2\overset{''}{.}00$ yr^{-1}
in the LHS, but these were generally found by other means. There is no
star in the LHS with such a motion fainter than m_R = 15.1.

Table 1 compares the luminosity function obtained from the LHS sample
with corresponding results from three other recent determinations. Here
our LHS sample has been restricted to stars classified as disk dwarfs on
the basis of photometric, astrometric, spectroscopic and kinematic
criteria. Hence, this sample should represent Population I stars in
disk orbits. The data in the third column are based on counting the
known stars within 5.2 pc of the sun in the northern 2/3 of the sky
(Dahn et al. 1986b). The fourth column gives the results obtained by
Gilmore & Reid (1983) from a color-selected (motion independent) sample
of stars. The fifth column shows the results of Wielen et al. (1983)
based on an analysis of an updated version of the Catalogue of Nearby
Stars (Gliese 1969; Gliese & Jahreiss 1979).

The tabulated results all generally indicate that a peak occurs in the
luminosity function (expressed in M_V) at +12-13. As discussed in
Dahn et al. (1986b), it is nonetheless apparent that the two nearby star
samples and this motion-selected (LHS) sample include more low luminosity stars with M_V > +15 than does the color-selected sample of
Gilmore and Reid. There are two logical reasons for this. First, since
the color sample is based on UK Schmidt plates exposed to deep limiting
magnitudes but over a limited solid angle of sky, the actual volume

Table 1. Luminosity Functions for Dwarf Stars (Units: 10^{+2}
x Number of Stars per Unit Interval of M_V per Cubic Parsec)

$M_V (\pm 0.5)$ Interval	LHS Sample	5.2 pc Sample	Gilmore & Reid	Wielen et al.
9.0	0.26±0.09		0.28±0.19	0.47±0.04
10.0	0.52±0.14	0.8±0.5	0.69±0.22	0.73±0.15
11.0	1.3 ±0.3	0.8±0.5	0.98±0.26	1.0 ±0.2
12.0	2.0 ±0.8	1.8±0.7	2.0 ±0.4	1.5 ±0.6
13.0	1.8 ±0.8	1.5±0.6	0.91±0.26	1.8 ±0.7
14.0	0.8 ±0.3	0.8±0.5	0.69±0.22	
15.0	0.8 ±0.3	1.8±0.7	0.35±0.16	1.3 ±0.4
16.0	0.5 ±0.3	1.0±0.5	0.16:	
17.0	0.7 ±0.4	1.3±0.6	<0.3:	0.9 ±0.3
18.0	0.2 ±0.1		<0.5:	
19.0	0.1 ±0.1		<0.2:	>0.05±0.03
20.0	0.1 ±0.1			
21.0				>0.01±0.01

surveyed for stars having M_V > +15 is actually rather small whereas the other studies include larger volumes. Secondly, the color sample identifies stars which are generally more distant and fainter in apparent magnitude but at a rather modest plate scale. Consequently, the many close binaries found in the nearby star samples would tend to be counted as single objects by Gilmore and Reid. This would have the effect of underestimating the faintest absolute magnitude bins, which contain significant fractions of close companions in the other samples.

The significant increase in the numbers of stars in the faint absolute magnitude bins does not lead to a drastic increase in the total mass density attributable to dwarf stars with M_V < +20. Formally, the LHS sample would add less than 0.02 $M_\odot$ pc^{-3} to the local mass density from stars with M_V >+15. However, this greater contribution would reduce significantly the amount required from "brown dwarfs," which might be fainter than M_V = +20. Thus, it is of special importance to identify whether a high luminosity "tail" of a distribution of substellar mass objects is detected at optical wavelengths.

3 PROPERTIES OF LOW LUMINOSITY STELLAR OBJECTS: METHODS FOR DETERMINING WHETHER THESE INCLUDE BROWN DWARFS

The lower mass limit at which stellar configurations are capable of sustaining thermonuclear reactions is not known with high accuracy from theoretical calculations. The principal uncertainty seems to be the amount and treatment of molecular opacity in the outer layers. For metal-poor stars, the mass limit may be as large as 0.10 $M_\odot$; for objects with metal and molecule rich surfaces, significant burning may persist for periods of one billion years for masses below 0.07 $M_\odot$, according to some calculations.

In addition, the conversion of an empirical M_V into a mass for an observed star requires knowing (1) the bolometric correction relation between M_V and ($L/L_\odot$), and (2) the mass-luminosity relation. In the absence of successful model atmosphere calculations to predict the emergent energy distribution, and given the uncertainties in the interior models, both relations are poorly known. The result is that we have a very poor working knowledge of the correct M_V (or range in M_V) at which stellar objects become substellar.

It is thus of considerable interest to develop empirical means of determining whether a given object of low luminosity or a group of such objects are substellar or not. Then we will want to apply such methods to an appropriate sample of objects which might be subjected to detailed observations.

3.1 Position in the H-R Diagram

Mike Bessell (1982) identified LHS 2397a as having a luminosity comparable to or possibly fainter than vB10 from spectrophotometric observations. Independently, Probst (1985; this conference)

obtained broadband red and infrared photometry which suggested that
LHS 2397a is cooler than vB10. A preliminary, unpublished USNO CCD
parallax (π_{abs} = 0.066 ± 0.002 m.e.) yields M_V = 18.5 ± 0.1, very
similar to the value of M_V = 18.73 ± 0.02 found for vB10 from the USNO
photographic parallax (Harrington et al. 1983).

While conversions of colors to temperatures and calculations of the
bolometric luminosities require atmospheric models which are not available, a redder color temperature for LHS 2397a combined with a similar
(or slightly brighter) M_V implies that its bolometric luminosity is
approximately the same or even brighter than that of vB10. The
simplest interpretation of their comparative positions in the H-R
diagram is that LHS 2397a has a smaller mass and a younger age than
vB10. However, one cannot rule out that the explanation involves simply
differing interior and/or atmospheric abundances. Moreover, the finding
does not demonstrate that LHS 2397a is a brown dwarf (or that vB10 is
not), but it does illustrate a method which, ultimately, may be applicable to well-observed individual objects.

3.2 Kinematics

It is well known that the velocity dispersion of a sample of
stars increases with mean age (cf., Wielen 1977). The effect is attributed to fluctuations in the gravitational field of the galactic disk,
possibly due to encounters with giant molecular clouds, an idea which
may have originated with the paper of Spitzer & Schwarzschild (1953).
Wielen (1977) shows that the velocity dispersion increases with
decreasing stellar mass (and increasing main-sequence lifetime) in an
apparently monotonic manner. Using the spectroscopically selected
McCormick K and M dwarfs, he demonstrates that the mean ages of these
stars are at least several billion years, since they show the largest
velocity dispersions.

Any sample of substellar mass objects which are capable of being visible
for only a short time must, therefore, have markedly smaller velocity
dispersions than the long-lived M dwarf stars. Poveda & Allen (1985)
recently attempted to identify a cutoff M_V value fainter than which
the objects show the expected smaller velocity dispersions. They in
fact suggested that this cutoff may be identified with the peak in the
empirical luminosity function near M_V ~ +12-13. Their study relied on
data from the Catalogue of Nearby Stars (Gliese 1969; Gliese &
Jahreiss 1979) which does not include data on many recently-discovered
low luminosity objects. The new radial velocities of Hartwick et al.
(1984) and especially Giampapa & Liebert (1986), combined with newly
derived distances and tangential velocities for the LHS sample, make it
possible to address this question more successfully. The sample is
still small and biased, but Giampapa and Liebert were able to reach the
following conclusions. First, the majority of stars at M_V ~ +15 to +16
(about ten) do show old disk kinematics, indicating a mean age of at
least several billion years. Likewise, there are several subdwarfs
with halo kinematics reaching M_V ~ +15. Thus we can conclude that the

empirical M_V limit for brown dwarfs is at least +16 for the Population I stars in the solar neighborhood, and +15 for stars with halo kinematics (which should be at least moderately metal poor). However, for M_V > +17 the situation is ambiguous and no useful conclusion is yet reachable. Many of the handful of well-studied objects show young disk kinematics, though such stars as vB8 and vB10 have indicated ages of a billion years or so.

3.3 Chromospheric activity
The surface activity level of a star having a convective envelope (a "solar type" star) is also known to correlate with age, though the primary dependence may be on the rotation rate. Differential rotation drives the magnetic dynamo which results in the dissipation of rotational energy via chromospheric and coronal activity. Unless the star is a close binary, the level of activity generally declines with age. Thus, such individual observables as the strength of central emission reversals of the Ca II H&K lines show a tight correlation with age for single stars (Wilson & Woolley 1970). Herbig (1985) shows that the same effect holds for Hα emission in F-G dwarfs and it has long been known that dwarf M stars in young clusters and groups generally show strong Hα emission.

We thus expect that a young, substellar mass population might also be characterized by strong Hα emission, and other indicators of chromo-spheric activity. Conversely, a sample of stars with a mean age of at least several billion years should not show a high incidence of such activity, unless the sample includes a large fraction of close binaries. Indeed, both the old disk stars near M_V ~ +15-16 and the halo stars near +15, cited in the previous section, uniformly showed no evidence of Hα emission in high dispersion (echelle) spectra.

On the other hand, most stars in the LHS sample fainter than M_V ~ +17 do show strong Hα emission. The sample is still too small for a useful conclusion to be reached, however. In particular, the otherwise pecul-iar object LHS 2924 near M_V = 19.6 shows at best a barely detectable Hα emission line.

REFERENCES

Bessell, M. (1982). Private communication.
Dahn, C.C., Liebert, J., Boeshaar, P.C. & Probst, R. (1986a). In preparation.
Dahn, C.C., Liebert, J. & Harrington, R.S. (1986b). LHS 292 and the luminosity function of the nearby M dwarfs. Astron. J. (in press).
Giampapa, M.S. & Liebert, J. (1986). High resolution Hα observations of M dwarf stars: Implications for stellar dynamo models and stellar kinematic properties at faint magnitudes. Astrophys. J. (in press).

Gilmore, G. & Reid, N. (1983). New light on faint stars-III. Galactic structure towards the South Pole and the Galactic thick disc. Mon. Not. Roy. astron. Soc.,202, 1025-47.

Gliese, W. (1969). Catalogue of Nearby Stars. Veroff. Astron. Rechen-Inst. Heidelberg No. 22.

Gliese, W. & Jahreiss, H. (1979). Nearby star data published 1969-1978. Astron. Astrophys. Suppl.,38, 423-48.

Harrington, R.S., Kallarakal, V.V. & Dahn, C.C. (1983). Astrometry of the low-luminosity stars vB8 and vB10. Astron. J.,88, 1038-9.

Hartwick, F.D.A., Cowley, A.P. & Mould, J.R. (1984). Studies of late-type dwarfs. VI. Identification of Population II main-sequence stars at M_V = +14. Astrophys. J.,286, 269-75.

Herbig, G.H. (1985). Chromospheric Hα emission in F8-G3 dwarfs, and its connection with the T Tauri stars. Astrophys. J.,289, 269-78.

Liebert, J., Boroson, T.A. & Giampapa, M.S. (1984). New spectrophotometry of the extremely cool proper motion star LHS 2924. Astrophys. J.,282, 758-62.

Luyten, W.J. (1979). LHS Catalogue. Minneapolis, Minn.: University of Minnesota Press.

Luyten, W.J. & La Bonte, A.E. (1973). The South Galactic Pole. Minneapolis, Minn.: University of Minnesota Press.

Monet, D.G. (1985). Private communication.

Monet, D.G. & Dahn, C.C. (1984). An updated trigonometric parallax for LHS 2924. Bull. Am. astron. Soc.,16, 1014.

Poveda, A. & Allen, C. (1985). Does the maximum of the stellar luminosity function correspond to a maximum of the mass spectrum? Preprint.

Probst, R.G. (1985). A photometric survey of the stars of largest H. Bull. Am. astron. Soc.,17, 553-4.

Probst, R.G. & Liebert, J. (1983). LHS 2924: A uniquely cool low-luminosity star with a peculiar energy distribution. Astrophys. J.,274, 245-51.

Schmidt, M. (1975). The mass of the Galactic Halo derived from the luminosity function of high-velocity stars. Astrophys. J., 202, 22-9.

Spitzer, L. & Schwarzschild, M. (1953). The possible influence of interstellar clouds on stellar velocities. II. Astrophys. J.,118, 106-12.

Wielen, R. (1977). The diffusion of stellar orbits derived from the observed age-dependence of the velocity dispersion. Astron. Astrophys.,60, 263-75.

Wielen, R., Jahreiss, H. & Krüger, R. (1983). The determination of the luminosity function of nearby stars. In The Nearby Stars and the Stellar Luminosity Function, IAU Colloq. No. 76, ed. A.G.D. Philip & A.R. Upgren, pp. 163-70, Schenectady, N.Y.: L. Davis Press, Inc.

Wilson, O. & Woolley, R. (1970). Calcium emission intensities as indicators of stellar age. Mon. Not. Roy. astron. Soc., 148, 463-75.

DISCUSSION

BOESHAAR: RG0050-2722 is quoted a being as intrinsically faint as LHS 2924, yet the spectrum and color data do not support this. Where does this M_V come from?

LIEBERT: McCarthy's viewgraph lists $M_V \sim$ +19.5, which may be based on a photometric parallax from Gilmore and Reid, although I remember only the claim that it was similar in colors to vB10 ($M_V \sim$ +18.7). My own spectrum (see Liebert & Ferguson, Mon. Not. Roy. Astron. Soc., 199, 29P-30P, 1982) is consistent with the latter claim.

BAHCALL: What about the claim of Gilmore and Reid that objects with $M_V \leq$ +19 can't make up the missing disk mass?

LIEBERT: Gilmore and Reid do not have a strong constraint on stars with $M_V \geq$ +16, although the shape of their function is well determined through the peak at $M_V \sim$ +13. Their two serious limitations, discussed in the text here, are (1) the small volume searched for low luminosity stars, and (2) the inability to detect close companion stars. Nonetheless, I do not think that we can account for the missing mass based on the numbers of objects between M_V = +16 and M_V = +19. I came to this meeting to learn whether it is possible that the remainder may be substellar!

THE SEARCH FOR LOW-MASS STELLAR COMPANIONS WITH THE HF PRECISION VELOCITY TECHNIQUE

Bruce Campbell
Dominion Astrophysical Observatory, Herzberg Institute of
Astrophysics, Victoria, B. C., Canada V8X 4M6

G. A. H. Walker
Department of Geophysics and Astronomy, University of
British Columbia, Vancouver, B. C., Canada V6T 1W5

Chris Pritchet & Barbara Long
Physics Department, University of Victoria, Victoria, B. C.,
Canada V8W 2Y2

Abstract. We have been observing 21 solar-like stars since 1980 with the hydrogen fluoride precision velocity technique at the Canada-France-Hawaii telescope. For spectra of sufficiently high signal-to-noise ratio ($\gtrsim$800:1) this method gives relative radial velocities with internal errors of $\sim$10 m s^{-1}. We have analyzed spectra from four stars to date, and three show no evidence of velocity variations with amplitudes greater than 20-25 m s^{-1}. This implies that these stars do not have companions of more than a few Jupiter masses in short period ($\lesssim$5 years) orbits. The fourth star, Procyon, shows velocity variations of $\sim$40 m s^{-1}, possibly related to the light variablity suspected of this star.

INTRODUCTION

There are two basic methods for detecting sub-stellar mass objects. The first is the direct approach - detecting photons emitted by the object itself, as with imagery, photometry, or speckle interferometry. In the second approach a star is used as a test particle to infer the presence of an unseen companion through its dynamical effects. Historically this second approach has been largely confined to astrometric techniques, with a number of candidates for brown dwarf companions coming from long-term astrometry programs (cf. Harrington 1985). However, there is a new generation of radial velocity techniques coming into use which are sufficiently sensitive that companions as small as Jupiter, or perhaps smaller, could be detected. (For a review of these new techniques, see Campbell & Walker 1985.) This opens the possibility of detecting low-mass companions in relatively short period orbits, since the velocity amplitude for a given mass companion varies as (separation)$^{-1/2}$. The astrometric perturbation, on the other hand, increases with the separation, and so these techniques are to a certain extent complementary. They also tend to be complementary in that astrometric programs have generally concentrated on faint stars (apparent and absolute), while precision radial velocity programs are usually confined to naked eye stars.

One of the new precision velocity methods is the hydrogen fluoride absorption cell technique proposed by Campbell & Walker (1979). We have been accumulating observations with this technique for the past five years at the Canada-France-Hawaii 3.6 m telescope. At the same time we have been developing the necessary software to analyze the spectra, and present in this paper the first results from the complete analysis package. The analysis of this type of data is by no means straightforward, since there are a number of subtle effects which can be ignored in conventional velocity measurements, but which require attention if systematic errors at the 10 m s^{-1} level, on a time scale of years, are to be eliminated. We therefore present along with the results, a complete description of the observing procedure and the reduction package.

THE HYDROGEN FLUORIDE ABSORPTION CELL TECHNIQUE

This technique has been previously described by Campbell & Walker (1979), Campbell et al. (1981), Yang et al. (1982), Campbell (1983), and Walker et al. (1984). It involves placing an absorption cell containing hydrogen fluoride gas ahead of the slit of a coudé spectrograph. Starlight passing through the cell then has absorption lines of HF superimposed, as shown in Figure 1. These lines provide precise wavelength fiducials which cannot be displaced relative to the stellar absorption lines by optical effects within the spectrograph. This is in contrast to conventional techniques, in which a separate beam of photons is used to provide wavelength calibration (comparison spectrum), and a variety of effects lead to systematic displacements in the focal plane of the spectrograph (Griffin & Griffin 1973). Hydrogen fluoride was chosen for this application because it produces strong lines for a cell length of only 90 cm, the lines are well spaced so that unblended stellar lines fall between, and there is no interference by lines of minor isotopes. (Fluorine has only one stable isotope, while bands of DF are displaced well away from the 8700 Å region). As well, the R-branch lines of the 3-0 band (Fig. 1) fall in a region free of telluric absorption lines, which are to be avoided for precision radial velocities (Caccin et al. 1985).

To ensure the purity of the hydrogen fluoride, the cell is baked at ~120° C and evacuated prior to each observing run. Fresh HF is distilled into the gas handling system every 6-8 months. Between runs the HF is saved by freezing with liquid nitrogen into a small sample cylinder. To avoid wavelength shifts, the cell temperature and pressure must be stabilized during observations. The temperature control is passive, with the observer adjusting heaters as necessary to maintain a temperature of about 100° C. The HF must be heated to about this temperature to avoid polymerization of the gas. The pressure of the HF is stabilized simply by immersing a reservoir of excess hydrogen fluoride in ice water (distilled). The gas pressure in the cell is then the vapor pressure of HF at 0° C, which is about 360 torr. A capacitance manometer attached to the cell shows that indeed the operating pressure has been constant to about 10 torr in five years of operation. In addition, we have found that the strengths of the HF

Figure 1. Spectra of β Vir with
(lower) and without (upper)
hydrogen fluoride lines
superimposed.

Figure 2. Relative displacement
of lines after convolution with
an instrumental profile
asymmetric by 610 m s⁻¹.

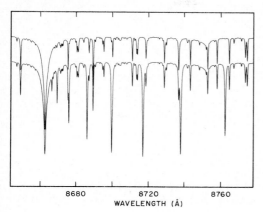

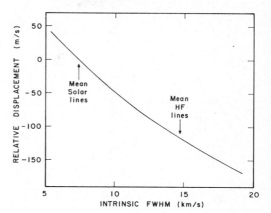

lines themselves provide a reliable check on the gas pressure and
temperature. Increasing pressure causes all lines to become stronger,
while changing the temperture changes the relative line strengths. This
enables measurement of these quantities <u>independent</u> of meters — which
could conceivably drift on a time scale of years — and therefore permits
reliable correction for pressure and temperature induced wavelength
shifts as calibrated by Campbell & Walker (1979) and Campbell (1983).

The observations have been made with the f/8.2 coudé spectrograph of the
Canada-France-Hawaii telescope on Mauna Kea. With the 830 groove/mm
mosaic grating this produces a dispersion of 4.75 Å mm⁻¹. From detailed
analysis of HF spectra we have found that the instrumental profile of
the spectrograph plus detector has full-width at half-maximum of about
33 microns, or about 0.16 Å, at 8700 Å. This is narrower than reported
previously by ourselves and other users of this spectrograph, since we
have now taken account of the intrinisic width of the measured lines.

The detector is a Reticon photodiode array with 1872 elements on 15
micron centers, 750 microns tall. While this device is noisy in
comparison to CCD's (readout noise ~400 electrons, Walker et al. 1985),
the exposures are to such a high level (generally ~1.5 X 10⁶ electrons
per pixel, for a signal-to-noise ratio of ~1000:1) that photon noise
dominates. Since Reticon arrays have high quantum efficiency in the
near-infrared (~70%, Vogt et al. 1977), as well as high geometric
stability, these devices are virtually ideal detectors for this
application. One disadvantage of these arrays, however, is their
sensitivity to cosmic rays, which limits exposure times to <2 hours.
This effect is somewhat more serious on Mauna Kea, where the cosmic ray
muon flux is about a factor of two greater than at sea level.

During each exposure of a star with HF superimposed, a chart recorder is
used to monitor the output of the spectrograph exposure meter.

Integration under the chart recorder tracing yields the mean time of the exposure, which is necessary to calculate the velocity correction to the solar system barycenter. This time must be known to ~2 minutes to ensure barycentric corrections accurate to ~5 m s^{-1}. For exposures of more than ~10 minutes, changes in transparency and guiding can mean that the mid-exposure time is simply not adequate in determining a precision radial velocity.

Each stellar exposure with HF is followed by four flat-field lamp exposures. An illumination system is used which produces the same beam shape as the starlight from the telescope. In addition, the axis of the flat-field beam is aligned with that of the starlight, to allow for the slight differences in coudé train collimation with declination and hour angle of each star observed. Experience has shown that such care in mimicking the beam of starlight is necessary to produce a reliable flat-field spectrum. We suspect that irregularities in the spectrograph optics, as well as the use of a Richardson image slicer (Richardson et al. 1971), contribute to this effect. Confirmation that this procedure yields high signal-to-noise spectra comes from the observed noise in the spectrum of the O-star ζ Ophiuchi obtained by Hobbs & Campbell (1982).

Spectra are obtained of hydrogen fluoride with the flat-field lamp three or four times each night. These spectra typically have signal-to-noise ratio >3000:1. These are used to monitor the gas temperature and pressure, and to measure changes in the instrumental profile of the spectrograph. As observing time permits, spectra are obtained of target stars without the HF superimposed. At least one such spectrum, preferably with signal-to-noise ratio >1000:1, is required to reduce the data.

REDUCTION PROCEDURE
Preparation of the spectra
The reduction of spectra in this program has been greatly facilitated by the use of the spectrophotometric reduction package RETICENT (Pritchet et al. 1982). This package enables manipulation of spectra both through its extensive kernel of commands, and through readily attached user-written routines for special purposes.

Each spectrum is first divided by flat-fields, then rectified with a third order polynomial fit to continuum points. Since it is important that all instrumental effects be removed from each spectrum consistently, the rectification is refined as follows. For each star a best spectrum taken without HF lines is chosen as the "reference" spectrum; let this spectrum be denoted S. Similarly, a reference spectrum of HF alone, H, was selected, generally from the same night as the reference stellar spectrum. Then for each stellar spectrum with the HF lines, SxH, the ratio

$$R = \frac{(SxH)}{(S)(H)},$$ (1)

should be 1.0 across the spectrum, provided that lines in S and H have been aligned with SxH. Of course there are generally glitches at the

locations of strong lines, which can be ignored. Between the lines the
ratio spectrum is in general not exactly 1.0 everywhere, and can consist
of two components. One is a random low frequency component, with
undulations over typically several hundred pixels, and amplitudes of
order 0.2%. The second is a sinsusoidal-like pattern with approximately
constant amplitude (<0.1%) across a spectrum, but with period
continuously varying from about 5 pixels at the blue end to about 300
pixels at the red end. This is presumably a fringing effect within the
Reticon array. The fringes have significant amplitude in only about 10%
of the spectra. Both of these components of the ratio spectrum, when
present, are fit with analytic functions, and the S̲x̲H̲ spectrum is
corrected by dividing by these functions. This procedure ensures that
each S̲x̲H̲ spectrum is rectified in precisely the same way relative to the
reference S̲ and H̲ spectra.

Line positions

Since the hydrogen fluoride lines are always blended with
stellar lines (at such high signal-to-noise ratio there are no truly
line-free regions in solar-type spectra!), and a few stellar lines are
blended with HF lines, it is necessary to cancel one set of lines (using
the reference spectrum) before measuring positions for the other. To
cancel the lines accurately it is generally necessary to shift and
stretch the reference spectra. (In other words, we allow for zero-
through third-order changes in the dispersion curve.) Where a strong
stellar line (>30% deep) is blended with an HF line, both are ignored in
aligning the spectra. To further check on the effect of stellar-HF
blends, a second cancellation is made with the reference spectrum
shifted by 0.04 pixels relative to its nominal (aligned) position. That
is, the reference spectrum is purposefully mis-aligned by an amount
equal to the typical error in absolute line positions. Relative line
positions (derived as described below) from this spectrum are compared
with those for the case of nominal alignment, to discern which lines are
sensitive to the alignment. In this way only lines significantly
affected by stellar-HF blends are identified and rejected from further
analysis. In practise, the HF lines are so strong that they are never
rejected by this criterion, while in general one or two stellar lines
per spectrum are rejected.

To determine the positions of the spectral lines we have used the difference technique of Fahlman & Glaspey (1973). This involves a series of
small shifts of a spectrum (ususally 0.1 pixel), and for each shift
subtracting the spectrum from the reference, and forming the sum of the
squares of residuals in the vicinity of each line. In the ideal case,
the summed squares versus shift form a parabola, the minimum of which
gives the position of the line in the object spectrum relative to the
reference. In practise, we fit a third-order polynomial to the residuals, to allow for line profile changes. Note that relative positions
are of primary importance here, since we are looking for velocity
changes. Absolute line positions are only used in aligning spectra.

Of several line position methods we have tested, that of Fahlman &
Glaspey (1973) has proved superior in minimizing random and systematic

errors. This technique can be used on asymmetric or blended lines (which are common in stellar spectra), and is not sensitive to either the summation end-points, or pixel-comb versus line position effects. As a general rule-of-thumb, we find that the method gives line positions with uncertainties of (signal-to-noise ratio in line)$^{-1}$ pixels.

The relative line positions so determined must be corrected for two subtle effects. The first results from slight changes in the instrumental profile of the spectrograph from run-to-run. If $\underline{S}$ is the intrinsic spectrum, and $\underline{I}$ is the instrumental profile, then the observed spectrum is the convolution of the two, $\underline{S*I}$. A theorem of convolutions states that the abscissae of the centers of gravity add (Bracewell 1965); that is

$$\langle x \rangle_{S*I} = \langle x \rangle_S + \langle x \rangle_I, \tag{2}$$

where
$$\langle x \rangle_F = \frac{\int_{-\infty}^{\infty} x\, F(x) dx}{\int_{-\infty}^{\infty} F(x) dx}. \tag{3}$$

This would imply that the relative positions of stellar and HF lines should not be affected by changes in the the instrumental profile. However, the positions of real spectral lines are determined by "integrals" (sums) over finite intervals, and so equation (2) does not strictly apply. To test for a possible effect of finite summation intervals, we have simulated spectral lines with realistic (Lorentzian) profiles, and convolved these with an asymmetric instrumental profile. To our surprise, we have found that with changes in instrumental asymmetry, the relative line position depends on the intinsic width of the line. This is illustrated in Figure 2. The effect is important here because the HF lines are on average about twice as wide as typical solar lines. Therefore a change in instrumental asymmetry could introduce a systematic velocity error. Changes in instrumental asymmetry as large as $\sim$250 m s^{-1} (defined as the difference in blue versus red half-intensity half-widths of the profile) are seen in our CFHT data, so that errors of $\sim$50 m s^{-1} might be introduced.

The second effect relates to blends of stellar and HF lines, and was discussed by Campbell & Walker (1985). Briefly, the cancellation of stellar lines by spectral division is not perfect because the observed spectra are convolved with the instrumental profile. Therefore

$$H' = \frac{(SxH)*I}{S*I} \tag{4}$$

is not the same as $\underline{H*I}$, unless $\underline{I}$ is very narrow compared to features in $\underline{S}$ and $\underline{H}$. Depending on the relative positions and strengths of stellar and HF lines, this effect can result in positions errors of several tens of meters per second (Campbell & Walker 1985).

To correct for both of these effects we first measure the instrumental profile width and asymmetry from HF spectra obtained each night. From measured positions, intrinsic strengths, and intrinsic widths of all stellar and HF lines we generate artificial spectra of the star+HF, reference star, and reference HF. These are then convolved with an instrumental profile, which is asymmetric only for the artificial star+HF spectrum. The degree of asymmetry is simply the difference between that of HF spectra of each night, and the reference HF. This set of artificial spectra is analysed in exactly the same way as the real spectra. Since the intrinsic positions of all stellar and HF lines are the same in the reference spectra and the star+HF spectrum, the resulting position differences are due only to the asymmetry and convolution effects. Thus the position differences from the artificial spectra form corrections for these effects, which are applied to the relative positions from the real spectra.

At the same time as the relative line positions are derived, additional information is extracted from the stellar lines. Relative to the reference spectrum we measure changes in equivalent widths and intrinsic asymmetries. The latter may be particularly useful in discerning changes in surface convection patterns due to stellar activity cycles, which could affect the apparent radial velocities (Dravins 1985). We also measure the intrinsic line widths, which could be useful in detecting changes in magnetic field strengths and surface filling factors. We note that our spectral resolution is not optimum for measuring these quantities, but that this is partially compensated by the very high signal-to-noise ratio of the data.

Radial velocities
To derive relative radial velocities from the line position data, we first fit a "dispersion" relation (ie. position difference versus absolute position) to the relative positions of the HF lines. This is used along with the stellar position differences to determine an initial estimate of the relative velocity. Then all stellar relative positions are adjusted to their "zero velocity" values, and the combination of these with the HF relative positions are used to generate another dispersion relation. The point is that we are utilizing the relative stellar line positions to constrain the higher than zero-order components of the dispersion relation. One further iteration of this procedure is normally sufficient to ensure convergence. Stellar lines more than 100 pixels outside the spectral range of the HF lines are not used in the velocity determination, but are used to constrain the ends of the dispersion relation.

In the calculation of velocities, and fitting dispersion relations, lines are weighted according to their strength. The weights are set to zero if either (a) they give discrepant velocities and are blended with lines of the opposite type, or (b) they are discrepant at more than the three sigma level. The latter rejection criterion we consider valid because the occasional cosmic ray event will occur in a line wing and produce a discrepant position. Such events will not in general be otherwise detectable because of the very high signal levels.

Relative line weights are further adjusted on examining the velocity residuals of each line for the complete set of spectra of a given star. Some lines tend to have larger scatter than expected from their strengths, possibly because they fall on the detector in regions of relatively large high-frequency components of the dispersion relation (Campbell et al. 1981). After all spectra are reduced, the line weights are then adjusted accordingly, and the reduction to velocities repeated. A third iteration produces slight additional changes. The internal errors of the final velocities are established from the weighted standard errors of the velocities of the typically 10-12 stellar lines of non-zero weight.

The final step is correction of the apparent velocities to the solar system barycenter. For this we use a barycentric correction program generated by Stephenson Yang, which utilizes the earth orbital routines of Stumpff (1980). The uncertainty in this correction is dominated by uncertainty in the mean exposure time, which generally amounts to less than 5 m s^{-1}.

RESULTS FOR FOUR STARS

The primary targets of the HF velocity program at CFHT are 21 F-K type dwarfs and subgiants not known to be close binaries. About half of these stars have now been observed more than 15 times, and of these we have performed full reductions for τ Ceti, Procyon, β Virginis, and β Comae. We consider the results for each of these in turn.

The velocities for τ Ceti (HR 509, G8V) from 22 spectra are shown in Figure 3. The points with parentheses indicate spectra with obvious faults - such as very poor spectrograph focus - and which can be ignored. In addition, there is a series of runs in mid-1981 (indicated by arrows in Figures 3-7) which produce velocities consistently high by

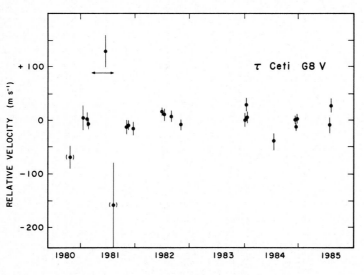

Figure 3. Relative velocities for τ Ceti.

~100 m s^{-1} in all four stars. At present, we do not know the origin of these discrepancies, and we clearly need to investigate this problem further by reducing more stars, and by examining in detail the affected spectra. For the moment we will ignore these points as well.

For the remaining 19 points in Figure 3 the velocities are remarkably constant. The mean internal error is about 11 m s^{-1}, while the standard deviation of the velocities is 14 m s^{-1}. We conclude from this that there is no evidence of a real velocity variation in τ Ceti.

The results for Procyon (HR 2943, F5IV-V) are shown in Figure 4. In this case there is an obvious velocity variation, which is due to a white dwarf companion. Figure 5 shows the velocity residuals after subtracting the orbital motion using the elements of Batten et al. (1978). In this case there do appear to be velocity variations at the level of ~40 m s^{-1}, since many points fall more than one sigma from the mean velocity. The variations appear to have been undersampled, but must have period longer than a few hours since no obvious variations are apparent in the time series observations of Campbell (1983). Perhaps these variations are related to the light variability suspected of Procyon (Hoffleit & Jaschek 1982).

The relative velocities for β Virginis (HR 4540, F9V) are plotted in Figure 6. Ignoring the spurious spectra and results from runs in mid-1981, we find that the point-to-point scatter in the velocitites is 18 m s^{-1}, while the average internal error is 10 m s^{-1}. We conclude that the velocity has remained constant over the five years of observation.

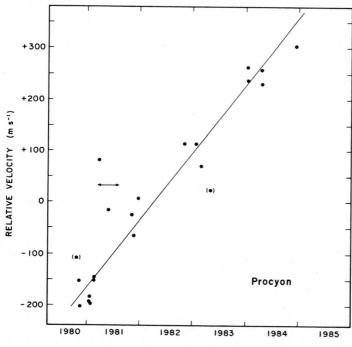

Figure 4. Relative velocitites for Procyon. Line shows expected velocity variation from spectroscopic orbit

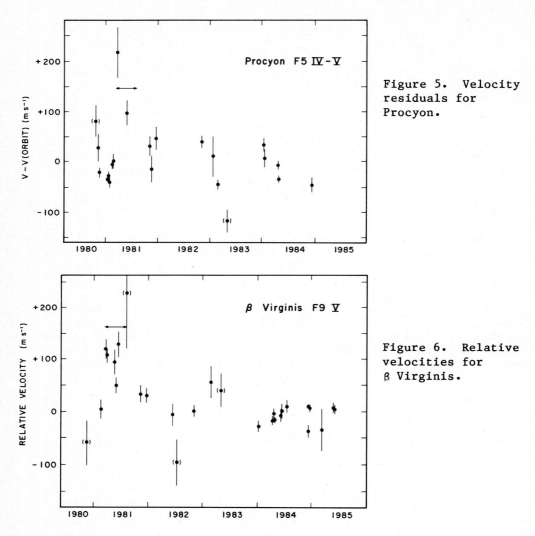

Figure 5. Velocity residuals for Procyon.

Figure 6. Relative velocities for β Virginis.

Finally, the results for β Comae (HR 4983, GOV) are shown in Figure 7. Again there is no sign of a real velocity variation, with mean internal errors of 14 m s^{-1}.

The results for these four stars are summarized in Table 1. The second column gives the weighted standard deviations of the velocities, with the spurious spectra and mid-1981 data omitted. The third column gives the mean internal errors. To estimate the upper limits on possible underlying variations we use the criterion

$$\text{limiting amplitude} = \sqrt{2} \times \text{external error.} \qquad (5)$$

This is a moderately conservative limit, since in the absence of noise

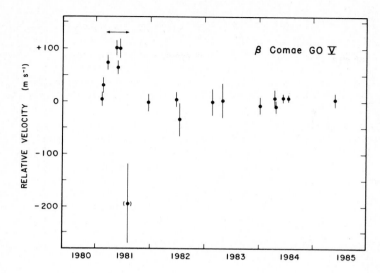

Figure 7. Relative velocities for β Comae

the r.m.s. variation of a sine wave is (amplitude)/√2. We use for the external errors the larger of the standard deviation in velocities, and the mean internal error. This yields the amplitude limits in the fourth column of Table 1. To establish the corresponding limits on masses of companions in orbits of 5 years or less (fifth column) we use

$$\frac{M_P}{M_J} \sin i = \left(\frac{\text{amplitude}}{16.6 \text{ m s}^{-1}}\right) \left(\frac{P_{yr}}{5}\right)^{1/3} \left(\frac{M_*}{M_\odot}\right)^{2/3}, \tag{6}$$

where M_J is the mass of Jupiter. While the individual limits have unknown orbital inclination factors, it is clear that companions of a few Jupiter masses cannot be present around most of these stars, unless they have orbital periods much greater than 5 years.

Table 1. HF radial velocity results for four stars.

Star	Standard deviation of velocities (m s^{-1})	Mean internal error (m s^{-1})	Maximum amplitude (m s^{-1})	Maximum $\frac{M_P}{M_J} \sin i$
τ Ceti	14	11	20	1.1
β Vir	18	10	25	1.6
β Com	11	14	19	1.2
Procyon	28	11	40	2.9

We conclude from this limited sample that companions somewhat more massive than Jupiter, whether brown dwarfs or planets, are not common close to solar-type stars. This result is perhaps not too surprising. Low-mass stellar or sub-stellar companions are not expected on observational grounds. In their survey of solar-like stars, Abt & Levy (1976) found that close binaries are predominantly of near unit mass ratio. This implies that these binaries are fission systems, and that proto-stellar clouds do not readily fragment into very dissimilar units.

On the other hand, giant planets presumably cannot form too close to a proto-star, since the circumstellar nebula must be cool enough to allow condensation of icy material. Predictions of how close they could form are not very precise, but Black & Scargle (1982) have suggested that massive planets may not occur in orbits of less than ~3 years period around solar-type stars. However, it is clear that as we extend our time base, and increase our sample of stars, we will be able to place increasingly interesting limits on the frequency of massive planets.

ACKNOWLEDGEMENTS

We thank Stevenson Yang and Chris Morbey for providing computer programs used in this work, and John Amor for his assistance with the observations and data reduction. We are most grateful to the staff of CFHT for their continuing assistance with the HF precision velocity program. Support for this work was provided by the National Research Council of Canada, and the Natural Sciences and Engineering Research Council of Canada.

REFERENCES

Abt, H. A., & Levy, S. G. (1976). Ap. J. Suppl., **30**, 273.
Batten, A. H., Fletcher, J. M., & Mann, P. J. (1978). Publ. D.A.O., **15**, No. 5.
Black, D. C., & Scargle, J. D. (1982). Ap. J., **263**, 854.
Bracewell, R. (1965). The Fourier Transform and Its Applications, p. 111. New York: McGraw-Hill.
Caccin, B., Cavallini, F., Ceppatelli, G., Righini, A., & Sambuco, A. M. (1985). Astron. Astrophys., **149**, 357.
Campbell, B. (1983). Publ. Ast. Soc. Pacific, **95**, 577.
Campbell, B. & Walker, G. A. H. (1979). Publ. Ast. Soc. Pacific, **91**, 540.
Campbell, B. & Walker, G. A. H. (1985). In Stellar Radial Velocities, I.A.U. Coll. 88, ed. A. G. Davis Philip & D. W. Latham, p. 5. Schenectady: L. Davis Press.
Campbell, B., Walker, G. A. H., Johnson, R., Lester, T., Yang, S., & Auman, J. (1981). Proc. S.P.I.E., **290**, 215.
Dravins, D. (1985). In Stellar Radial Velocities, I.A.U. Coll. 88, ed. A. G. Davis Philip & D. W. Latham, p. 311. Schenectady: L. Davis Press.
Fahlman, G. G., & Glaspey, J. W. (1973). In Astronomical Observations with Television-Type Sensors, ed. J. W. Glaspey & G. A. H. Walker, p 347. Vancouver: University of British Columbia.
Griffin, R., & Griffin, R. (1973). M.N.R.A.S., **162**, 243.

Harrington, R. S. (1985). These proceedings.

Hobbs, L. M., & Campbell, B. (1982). Ap. J., **254**, 108.

Hoffleit, D. & Jaschek, C. (1982). The Bright Star Catalogue. New Haven: Yale University Observatory.

Pritchet, C. J., Mochnacki, S., & Yang, S. (1982). Publ. Ast. Soc. Pacific, **94**, 733.

Richardson, E. H., Brealey, G. A., & Dancy, R. (1971). Publ.D.A.O., **14**, 1.

Stumpff, P. (1980). Astron. Astrophys. Suppl., **41**, 1.

Vogt, S. S., Tull, R. G., & Kelton, P. (1977). Appl. Optics, **17**, 574.

Walker, G. A. H., Amor, J., Yang, S., & Campbell, B. (1984). In Calibration of Fundamental Stellar Quantities, I.A.U. Symp. 111, ed. D. S. Hayes, L. E. Pasinetti, & A. G. D. Philip, p. 589. Dordrecht: Reidel.

Walker, G. A. H., Johnson, R., & Yang, S. (1985). Advances in Electronics and Electron Physics, **64A**, 213.

Yang, S., Walker, G. A. H., Fahlman, G. G., & Campbell, B. (1982). Publ. Ast. Soc. Pacific, **94**, 317.

DISCUSSION

J. TARTER: Will you continue monitoring the three stars on which you have null results in order to investigate the possible existence of quasi-periodic fluctutations in the stellar spectra at 10 m s^{-1}?

B. CAMPBELL: Yes we will, especially as we have been given some assurances of continued observing time at CFHT. In addition, we measure many parameters from our spectra which could reveal whether any velocity changes found are due to effects other than motion of the stellar center of mass.

N. G. ROMAN: (1) What kind of limits can be placed on systems with two or three Jupiter-mass companions? (2) Have you seen any line shape or line ratio variations which might confirm that Procyon is a δ Scuti star?

B. CAMPBELL: (1) Presumably such systems would have very dissimilar periods, and so we should be able to disentangle the variations, especially as the velocity amplitudes decrease with increasing period for a given mass planet. In the case of the solar system one planet very much dominates. Jupiter produces a reflex motion in the sun of 12.5 m s^{-1} amplitude, while the next largest effect is due to Saturn, which produces variations of only ~2.5 m s^{-1}. (2) No, we have not yet examined the data for the line profiles of Procyon.

D. DEMING: When you transform your velocities to the solar system barycenter you will remove the Doppler reflex of the sun due to Jupiter, because the earth largely shares this reflex. Have you tried transforming your data to a heliocentric frame? In this case you may be able to see, for stars near the ecliptic plane, a drift in velocity due to Jupiter.

B. CAMPBELL: No, we have not looked for this effect.

A DYNAMICAL SEARCH FOR SUB-STELLAR OBJECTS

G.W. Marcy, V. Lindsay, J. Bergengren, D. Moore
San Francisco State University

Abstract. Accurate radial velocities of 65 M dwarfs have
been obtained monthly, enabling detection of companions down
to 0.005 M_0. Only one star exhibits velocity variations
consistent with a sub-stellar companion.

INTRODUCTION

Among the efforts to detect sub-stellar objects, photometric
techniques (e.g., Boeshaar & Tyson 1985, and McCarthy 1984) have provided evidence that their number density may be less than that estimated from extrapolation of the lower main-sequence mass function.
Dynamical techniques, using astrometric or radial velocity measurements, permit direct mass estimates to be made, but tend to be slower.
Here we describe the results of a velocity approach which is fast,
enabling large numbers of targets to be surveyed.

A comparison of possible techniques for detecting sub-stellar companions to stars is shown in Figure 1, which shows the parameter space
of companion mass and orbital radius. The upper-right region is that
in which conventional ground-based astrometry is capable of detecting
sub-stellar companions. We have assumed a distance of 10 pc to the
system and a minimum detectable perturbation amplitude of 0.01 arcsec
for a convincing detection.

Also shown in Figure 1 is the detectability curve for the several
"precision-velocity" programs which are currently, or soon to be under
way (Campbell and Walker, 1984; McMillan et al.; and Cochran et al.).
We have taken 10 meters/sec as the total velocity error for these programs (the desired accuracy). The left boundary at 2 a.u. represents
the estimated minimum orbital radius of a low-mass, sub-stellar object
around a G2 dwarf (Black and Scargle, 1982).

Neither of the above approaches is designed to detect short-period
objects (P < 5 yrs), which is unfortunate since two proposed scenarios
for the formation of sub-stellar companions may favor small orbital
radii. In particular, formation by fragmentation of the primary star
may favor small semi-major axes (see for example, the G dwarf binary
statistics of Abt and Levy 1976). Second, formation within a

Figure 1. Detectability, at 10 pc, of sub-stellar companions by
different techniques, in the parameter plane of semi-major axis
vs. mass of companion. Note that the approach described here is
efficient at detecting sub-stellar companions within 2 a.u.
For astrometry and the present work, a 0.3 $M_{\odot}$ primary star was
assumed, while for the "precision velocity" work, a G2 dwarf
primary was assumed.

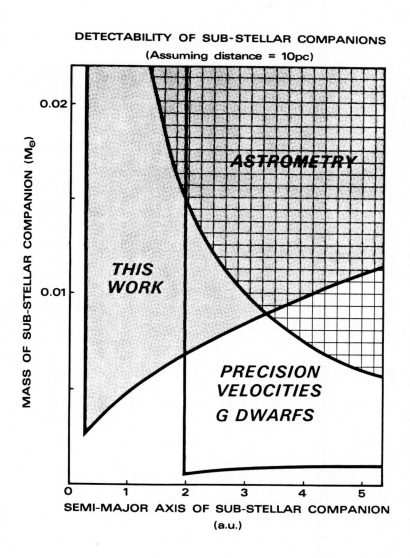

Fig. 1

protostellar disk is expected theoretically to yield the more massive
objects closer in, as evidenced by the distribution of masses among the
Jovian planets in our solar system. Figure 1 shows the region of
detectability using the adopted radial velocity approach in this work.
Sub-stellar companions with masses of 0.005 $M_\odot$ in the inner 1 a.u.
around an M dwarf primary are detectable, as described below.

The Approach: Reflex-Motion of M-Dwarfs
To detect sub-stellar companions of the lowest possible
mass, the optimal target primaries are dwarf M stars, because of their
low inertia (M = 0.3 $M_\odot$), their rich spectra (ideal for cross-
correlation techniques), and their low luminosities. The low luminosities reduce the temperature of the envelope of a sub-stellar
object, thereby inhibiting the escape of light-element constituents.
The low luminosities will also aid in the survival in ice grains,
possibly important in the formation of objects in protostellar disks
(c.f., Black and Scargle, 1982).

We have carried out a series of accurate velocity measurements of M
dwarfs during the past two years. The expected magnitude of the reflex
velocity of the primary is calculated with the usual expression for the
two-body problem:

$$K = 0.52 \ (1-e)^{-1/2} \ \frac{M_2/10M_{Jup}}{[(M_1+M_2)/0.3M_\odot]} \ a^{-1/2} \quad (km/s) \qquad (1)$$

where K_1 is the amplitude of the observed velocity variations, e is the
eccentricity, M_1 and M_2 are the masses of the primary and companion,
respectively, and a is the semi-major axis of the relative orbit in
a.u.'s.

The equation is expressed in canonical units to show that a 10
Jupiter-mass companion in circular orbit of radius 1 a.u. around an M
dwarf of mass 0.3$M_\odot$ (dM4) yields an amplitude of 0.52 km/s. (The peak-
to-peak variation is 1.04 km/s.) Of course, the observed velocity will
be decreased by sin i, the sine of the orbital inclination. Equation
(1) shows that companions having masses from 0.1 $M_\odot$ down to 10 M_{Jup}
orbiting M4 dwarfs within a few astronomical units produce reflex
velocities of at least several tenths of one km/s.

The Velocity Technique
The coudé spectrograph of the Mount Wilson 100-inch telescope was employed with a new grating to yield a reciprocal dispersion
of 1.0 Å/mm. The intensified reticon detector (Shectman, 1981) was
used, yielding 500 m/s per pixel. Several significant modifications
were made to the spectrograph to maximize stability of the optical system. The Doppler shift of each stellar spectrum is determined relative
to a suitable template spectrum via a cross-correlation analysis, and

nightly corrections are determined from the program stars themselves to register the zero point of the velocity measurements.

Observations over the course of two years show that the total error is about 200 m/s, resulting from the signal-to-noise ratio of the spectra (about 7:1) and imperfect guiding of the stellar image on the slit of the spectrograph. The magnitude of this effect depends on the seeing, with excellent seeing being the worst case (!) - a common problem at Mount Wilson. Future observations will be made at Lick Observatory.

Progress to Date
For the past two years a velocity program has been carried out using the techniques described above. An unbiased sample of 65 dwarfs of spectral types M2 to M5, were chosen from the list of Joy and Abt (1974). All are brighter than V=11.5, and are thought to be single (Wilson 1967, Gliese 1969): that is, known close binaries were excluded. Velocity measurements were made once per month for each star, when observable.

A typical set of results is shown in Figure 2, this for Gliese 649, showing a plot of measured velocity versus Julian day. The full verti- cal scale is 2.0 km/s, and the bar placed at the lower left has length 250 m/s. The observed standard deviation is 170 m/s, with no obvious periodicity present. It is worth recalling (from eq. 1) that a sub- stellar object, for example, of 10 Jupiter-masses orbiting at 1 a.u. would induce a velocity amplitude of 520 m/s x sin i. Assuming the average value, <sin i> = 0.79, these data suggest that it is unlikely that a sub-stellar object greater than 10 M_{Jup} exists within 1 a.u. of Gliese 649.

This case is representative of the sensitivity to sub-stellar com- panions that is achieved for all 65 program stars, with the limited data in hand. In particular, companions in a decade range of mass, from 0.1 solar masses to 0.01 solar masses are detectable, should they exist anywhere within the inner few a.u.'s of the target star.

One interesting case is that of Gliese 699, Barnard's star. Its stan- dard deviation of 240 m/s is slightly larger than average, but an F- test strongly suggests that this scatter cannot be considered signifi- cant relative to that of the other program stars. Of course, Van de Kamp's (1977) long-debated 11.7-year astrometric periodicity for Barnard's star would not be expected to show up after only 1 1/2 years of measurements. Thus these data of limited duration argue only against the presence of sub-stellar companions having periods less than about 2 years.

Note that because of the nature of Keplerian motion, orbits having periods less than the data sampling time of one month are nonetheless detectable. For example, a companion of mass 0.01 M_0 with an orbital period of only 20 days would induce a reflex velocity in the primary star of 1.6 km/s - an 8σ effect.

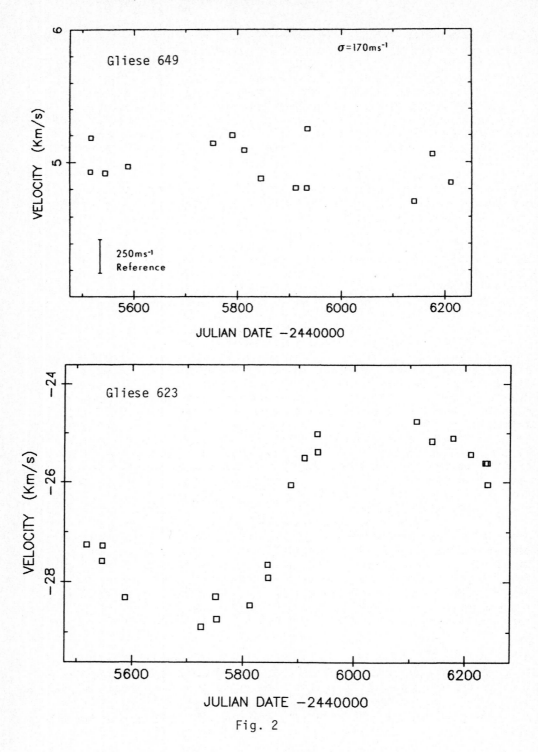

Fig. 2

Despite the broad range of sensitivity in orbital period and companion mass, few of the program stars exhibit any evidence of periodicity. Five of the 65 stars exhibit velocity scatter slightly over 300 m/s. However, an F-test reveals that one would expect a few stars to have 50% higher standard deviations than the average, given the number of observations and the estimated external error of 200 m/s.

Among the 65 M dwarfs, only two showed convincing velocity variability. Gliese 570 B exhibits a velocity amplitude of 6 km/s and period of about 300 days, and thus has a stellar companion. More interesting is Gliese 623 (figure 2) for which smooth velocity variations are seen with a period of approximately 2.2 years and amplitude of 2.0 km/s. Thus, the companion has a minimum mass of about 0.055 $M_\odot$. Assuming the average value for orbital inclination, <sin i> = 0.79, the companion mass is 0.070 $M_\odot$.

Interestingly, a companion to Gls 623 was already known from astrometry (Lippincott and Borgman, 1978) and from speckle inter-ferometry (McCarthy, 1984). However, the astrometric period of 3.7 years is so different from the velocity period of 2.2 years that the possibility of two companions is raised. The stability of two such companions is, however, doubtful, as the primary is no more than 10 times more massive than either.

Finally, in the sample of 65 M dwarfs, two exhibit inconclusive evidence of velocity variations. In both cases (Gls 70 and Gls 15A), the velocities seem to vary in a smooth way, not inconsistent with periodic changes. But the amplitudes are approximately equal to the error in the measurements. More measurements are needed to verify the possible variability in these stars. However, the indicated velocity amplitudes are so low in these cases that confirmed variability would imply the existence of extremely low mass sub-stellar objects.

DISCUSSION

Considerable information about the prevalence of sub-stellar companions may be extracted from the velocities of the 65 sample stars. Three stars were found to be double-lined spectroscopic binaries, and thus have stellar companions. Of the remaining 62 stars only 2, namely Gls 570B and Gls 623 mentioned above, were found to exhibit convincing velocity variations. Only the latter has orbital elements consistent with a possible sub-stellar companion, the minimum mass being below 0.08 $M_\odot$.

Thus of 62 M dwarfs, only one possible sub-stellar companion was detected above the minimum detectable mass of about 0.005 $M_\odot$ (see figure 1). Of course, the duration of the project to date limits the detectable orbital periods at about 2 years. Therefore one is limited to conclude that among the 62 single M dwarfs in this unbiased sample, at most one has a sub-stellar companion above 0.005 $M_\odot$ within one astronomical unit. This presumes that orbital inclinations are not extreme.

For greater orbital radii, one complete orbit would not have been completed during the course of the project. Nonetheless, one can calculate the minimum possible velocity variation expected from such orbits after only 2 years. Such an approach leads to the conclusion that for orbital periods less than, for example, 4 years, sub-stellar companions of mass greater than $0.020 M_\odot$ would have been detected. No such objects were seen, with the possible exception of Gls 623.

The qualitative conclusion seems clear, namely that sub-stellar objects are extremely rare, at least within a few astronomical units of M dwarfs.

REFERENCES

Abt, H. A. and Levy, S. G. (1976). Ap. J. Suppl., 30, 273.
Black, D. C. and Scargle, J. D. (1982). Ap. J., 263. 854.
Boeshaar, P. C. and Tyson, J. A. (1985). Astron. J., 90, 817.
Campbell, B. and Walker, G. A. H. (1984). I. A. U. Colloq. #88, "Stellar Radial Velocities."
Gliese, W. (1969). "Catalogue of Nearby Stars," Astronomischen Rechen-institut Heidelberg, No. 22.
Joy, A. H. and Abt, H. A. (1974). Ap. J. Suppl., 28, 1.
Lippincott, S. L. and Borgman, E. R. (1978). Pub. A. S. P., 90, 226.
McCarthy, D. W. (1984) in I. A. U. Colloq. No. 76.
Shectman, S. (1981). Carnegie Institution of Washington Year Book, 1980, p. 586.
Van de Kamp, P. (1977). Vistas in Astronomy, 20, 501.
Wilson, O. C. (1967). Astron. J., 72, 905.

DISCUSSION

McCarthy: The astrometric orbit of Gls 623 has a low amplitude. Is it possible that a single companion can explain both astrometric and radial velocity data within the errors?

Marcy: It seems unlikely. The astrometric and radial velocity periods of 3.7 and 2.2 years, respectively, are both determined to within about 0.2 years.

Tarter: What mass do you get for the radial-velocity companion of Gls 623 if you assume that it has the same orbital inclination as the astrometric companion?

Marcy: You get a mass of about $0.1 M_\odot$. But since the eccentricities determined for the two are found to be so different, it may not be reasonable to assume that the inclinations are the same.

Roman: Is it possible to see double lines in those stars that Harrington suspects are astrometric binaries? If so, can you determine whether his "comparable-mass" solution or his "large-mass-ratio" solution is correct?

Marcy: Unfortunately, there is very little overlap in our two samples. For those stars in common, a difficulty arises in that astrometric studies are preferentially sensitive to large semi-major axes, while radial velocity approaches are more sensitive to close companions. In principle, the two approaches are complimentary, and are able to resolve those ambiguities.

THE CONVECTIVE NOISE FLOOR FOR THE SPECTROSCOPIC DETECTION OF LOW MASS COMPANIONS TO SOLAR TYPE STARS

D. Deming*, F. Espenak and D. E. Jennings
Planetary Systems Branch, Code 693,
Goddard Space Flight Center, Greenbelt, MD   20771

J. W. Brault
National Solar Observatory
950 N. Cherry Avenue, Tucson, AZ   85726

Abstract. Low mass companions to solar type stars can, in principle, be detected by measuring the small changes which result in the radial velocity of the parent star. A possible limitation of this method is that changes in the properties of stellar convection may be confused with the Doppler reflex due to a low mass companion. We define the threshold mass for unambiguous spectroscopic detection to occur when the maximum acceleration in the stellar radial velocity, due to the Doppler reflex of the companion, exceeds the apparent acceleration produced by changes in convection. We have measured an apparent acceleration in integrated sunlight, using near infrared Fourier transform spectroscopy, obtaining 11 meters sec^{-1} year^{-1} in 1983. We attribute this drift in the apparent solar velocity to a lessening in the magnetic inhibition of granular convection, as solar minimum approaches. On this basis we estimate the threshold mass for spectroscopic detection of companions to a one solar mass star as:

$$M > 0.07\ r^2/\sin i \qquad \text{Jupiter masses,}$$

where r is the orbital radius for the companion in astronomical units. In the short period realm, where spectroscopic techniques are most complementary to astrometric efforts, this yields a detection threshold below one Jupiter mass.

Introduction. A promising method for the detection of sub-stellar mass objects is to infer their presence as unseen companions in binary systems. This can be accomplished by measuring the acceleration of the parent star, and using momentum conservation to infer the presence and orbital characteristics of the low mass companions. The acceleration components in the plane of the sky are

---

*Guest investigator, National Solar Observatory, a division of the National Optical Astronomy Observatories which is operated by the Association of Universities for Research in Astronomy, under contract with the National Science Foundation.

measured using astrometric techniques. The radial component of stellar
acceleration can be measured using spectroscopic radial velocity
techniques (Serkowski 1976, Campbell and Walker 1979, McMillan et al.
1985). Radial acceleration is produced only when $|\sin i| > 0$, where i
is the inclination of the orbital plane to the plane of the sky.
However, radial accelerations are largest for stars having companions
in short period orbits, where the astrometric measurement is most
difficult. Moreover the magnitude of the Doppler reflex produced by
the companion is independent of the distance to the star, allowing a
diverse sample of stars to be searched using radial velocity
techniques. For these reasons, the radial velocity technique is quite
complementary to astrometric efforts, and is being vigorously pursued
by several research groups.

The detection of low mass companions using the radial velocity
technique requires a very stable spectrometer. In addition to the
instrumental requirements, there is also the question of whether
stellar velocity fields are sufficiently stable over periods of years
to allow the unambiguous detection of low mass companions. If we use
the Sun as a representative star, we can ask to what extent there are
intrinsic long term changes in the radial velocity of integrated
sunlight. After removal of the gravitational redshift, solar lines are
found to be systematically blueshifted relative to laboratory
wavelengths. This is attributed to granular convection, in which hot,
rising, elements dominate over colder descending regions in the
integrated light spectrum (Dravins 1982). The nature and magnitude of
long term changes can be addressed in four ways:

(1) The morphology of surface granulation (i.e. number of granules,
their thermal contrast, etc.) can be studied as a function of time in
the solar cycle, and the expected effect on the apparent velocity of
integrated sunlight can be calculated. On this basis, Dravins (1985)
has estimated a 30 meter $\sec^{-1}$ change over the solar cycle.

(2) The blueward asymmetry of absorption line profiles can be studied
as a function of position on the solar disk. Livingston (1982) has
found that these blueward asymmetries are larger in non-magnetic than
in magnetic regions. Since regions of high surface magnetism are less
common at solar minimum, this would imply that line centroids will show
greater blueshifts at the minimum of the solar activity cycle.

(3) Robert Howard and his colleagues at Mt. Wilson have obtained
velocity maps of the solar surface in connection with solar rotation
studies. These data were taken with an unstabilized spectrograph, but
they are very extensive, covering more than two 11-year solar cycles.
Snodgrass (1984) has recently discussed these results, and has noted
that the data indicate an effect over the solar cycle. The sense of
the effect is of a greater blueshift at solar minimum. Snodgrass
remarks that this effect is lessened if regions of high surface
magnetism are excluded from the data.

(4) The centroids, or line bisectors, of absorption lines in integrated light can be monitored by direct observation with a very stable spectrometer. In this paper we report first-epoch results from such a program.

Observations. We obtained observations of integrated sunlight on six occasions in 1983, using the McMath Fourier transform spectrometer (FTS) on Kitt Peak. We observed the solar spectrum in the region from 4000 to 4800 cm^{-1}, at a spectral resolution of 0.012 cm^{-1}. The McMath FTS is a dual cat's eye Michelson interferometer, operated in vacuum. It produces spectra of a quality sufficient for work involving laboratory frequency standards. We observed in the near infrared because we could then calibrate our spectra to great precision using an absorption cell of N_2O at 20 Torr. The N_2O cell was placed in front of the entrance aperture of the FTS, and it completely filled the FTS field of view, which extended to twice the angular diameter of the Sun. Absolute vacuum wavenumbers for the N_2O lines were determined by one of us (J.W.B.) by measuring the N_2O lines relative to the P(7) line of methane. The N_2O rest frequencies are known to < 1 meter sec^{-1} in equivalent velocity, and the signal-to-noise ratio in our solar spectra ($\sim$ 3000) allowed us to calibrate the frequency scale to 1 meter sec^{-1} precision. We measured 16 lines in the $\Delta V = 2$ sequence of $^{12}C\ ^{16}O$, and 3 weak atomic lines. Weak lines in this spectral region are formed deep in the solar atmosphere, where the effects of convection should be especially prominent. Portions of a typical spectrum are shown in Figure 1. We obtained the solar velocity by fitting gaussians to the

Figure 1. Sample spectra in the region of the N_2O lines (top) and the solar CO region (bottom).

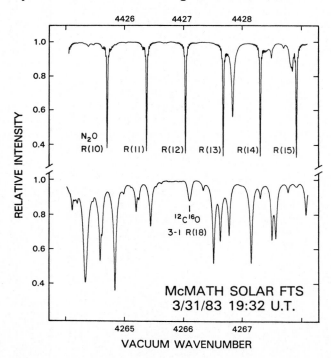

profiles of the solar lines. We found that these weak solar absorption lines were sufficiently symmetric that the line centroid was a reasonably meaningful concept. Our results for the integrated light velocity are absolute; rest frequencies for the $^{12}C\ ^{16}O$ lines were derived from the molecular constants of Mantz and Maillard (1974).

Although our spectra can be calibrated at the 1 meter sec^{-1} level, there are other sources of error which limit the accuracy of our measurements. We anticipated that imperfect optical integration over the solar disk would be the principal source of error. Since the solar rotation velocity varies by ± 1800 meters sec^{-1} across the disk, small angular variations in optical efficiency can lead to serious errors in the integrated light velocity. However the solar orientation rotates with respect to the FTS and telescope optic axes, with a 24 hour period. This modulated the systematic velocity errors as a function of hour angle, and allowed us to identify, model, and remove them. We found two principal sources of systematic error. At the lowest solar declinations differential transmission of the terrestrial atmosphere produces a velocity error (e.g. see Grec et al. 1977). This effect is illustrated on Figure 2, using December 1983 data. At easterly hour angles the approaching limb of the Sun is at larger airmass than the receding limb, leading to a spuriously large radial velocity, and vice-versa at westward hour angles. We modeled and removed the effect using a one free parameter model.

Figure 2. The effect of differential atmospheric transmission on the velocity of integrated sunlight, illustrated using data taken in December 1983. The velocity scale is absolute, but the solar gravitational redshift has not been removed.

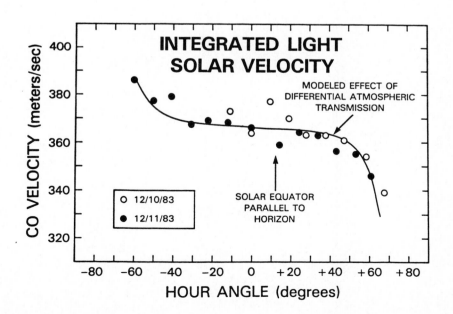

When the Sun is at positive declinations, a second source of systematic
error occurs due to foreshortening of the McMath heliostat. This
produces a sinusoidal velocity error with a 24 hour period and 60 meter
sec^{-1} amplitude. This is an effect of the telescope, not of the FTS.
It occurs because the rotation axis of the McMath heliostat is not
coincident with the telescope optic axis, leading to an hour angle
dependent vignetting. This is shown on Figure 3, which compares the
data for March and April 1983 with December 1983. The vignetting due
to the heliostat is a very reproducible effect. The April data, for
example, were taken by a different observer than for the March data
after a complete shutdown of the telescope and FTS. The velocity
error, however, is essentially identical for the March and April data.
The zero-crossing of the error curve gives the absolute velocity of
integrated sunlight; values for individual days are shown by dash marks
on the figure. It can be seen that the March/April values are quite
close to the velocity observed nine months later, in December 1983.
The fact that removal of two quite different sources of systematic
error give results in such good agreement is very convincing.

> Figure 3. Velocity results for March/April and December
> 1983, shown as a function of hour angle. All of the data
> have been corrected for the effect of differential
> atmospheric transmission. The two points enclosed in
> brackets were not used in the analysis. Daily values for
> the CO velocity are shown by horizontal lines, labeled by
> date.

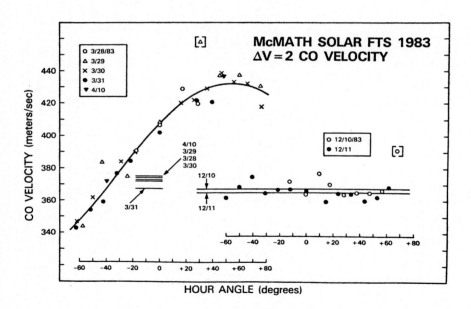

Discussion. Figure 4 summarizes our data for 1983. At the
top we plot the integrated light velocity with each point representing

an entire observing run. The size of the error bars depends strongly upon whether we could adequately define and remove the systematic effects. Two epochs have very large errors. The December 1982 data were taken through cirrus clouds, which produce large, uncorrectable, perturbations in the integrated light velocity. In July 1983 we were not able to obtain data over a sufficient range of hour angles to define and remove the effect of heliostat vignetting. For the remaining four epochs, the systematic errors could be conclusively removed. These results are expanded in the lower portion of Figure 4, where each point represents a single day's data, ranging from two to twelve spectra. Based on the spatially resolved studies cited earlier,

Figure 4. Results for the CO velocity of integrated sunlight in 1983. The top portion of the figure plots our data at all epochs, and the filled circles represent data where the systematic errors could be conclusively removed. The better data are re-plotted in the lower portion of the figure, on an expanded velocity scale. Values for individual days are shown, each point representing from two to twelve spectra.

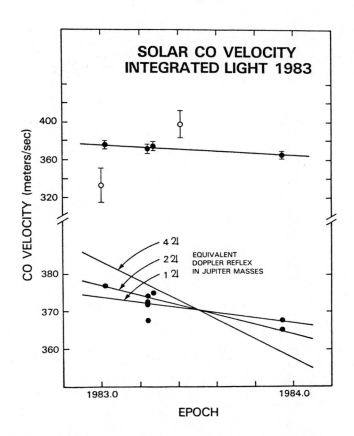

it is reasonable to expect a velocity variation which is sinusoidal, and in phase with the solar cycle. Over the one year interval spanned by our 1983 data such a variation would appear as an approximately linear drift toward a greater convective blueshift, as solar minimum approaches. Since it happens that the orbital period of Jupiter is close to the period of the 11-year solar cycle, it is interesting to compare the data to the equivalent Doppler reflex of Jovian mass companions with an 11-year orbital period. This comparison is also shown on Figure 4, where it can be seen that the data are most consistent with an effect of order two Jupiter masses. The sense and approximate magnitude of the effect is consistent with the expectations voiced by Dravins (1985). We define the threshold mass for unambiguous detection of a low mass companion to occur when the acceleration in the stellar radial velocity produced by the companion exceeds the apparent acceleration produced by changes in convection. We take the latter to be 11 meters sec^{-1} $year^{-1}$ based on our Figure 4 data. On this basis we estimate the detection threshold for spectroscopic detection of companions to a one solar mass star as:

$$M > 0.07 \; r^2/sin \; i \qquad \text{Jupiter masses,} \qquad (1.1)$$

where r is the orbital radius of the companion in astronomical units. In the short period realm (P $\sim$ few years) this yields a detection threshold well below one Jupiter mass. Our measured acceleration of 11 meters sec^{-1} $year^{-1}$ has a statistical error of order $\pm$ 4 meters sec^{-1} $year^{-1}$, so the intrinsic limits for detection of planetary mass companions are still somewhat uncertain. However brown dwarfs such as VB8 B are certainly well above the threshold for spectroscopic detection.

References.

Campbell, B. and Walker, G.A.H. (1979). Pub. Astr. Soc. Pac. 91, 540-545.

Dravins, D. (1982). Ann. Rev. Astron. Astrophys. 20, 61-89.

Dravins, D. (1985). In "Stellar Radial Velocities", IAU Colloquium No. 88, A. G. Davis Philip and D. W. Latham (eds.), L. Davis Press, Schnectady, N.Y., pp. 311-320.

Grec, G., Fossat, E., Brandt, P. and Deubner, F. L. (1977). Astron. Astrophys. 77, 347-350.

Livingston, W. C. (1982). Nature 297, 208-209.

Mantz, A. W. and Maillard, J. P. (1974). J. Molec. Spectros. 53, 466-478.

McMillan, R. S., Smith, P. H., Frecker, J. E., Merline, W. J. and Perry, M. L. (1985). In "Stellar Radial Velocities", IAU Colloquium No. 88, A. G. Davis Philip and D. W. Latham (eds.), L. Davis Press, Schnectady, N.Y., pp. 63-86.

Serkowski, K. (1976). Icarus 27, 13-24.

Snodgrass, H. B. (1984). Solar Physics 94, 13-31.

DISCUSSION:

CAMPBELL: 1) You are placing limits on minimum detectable companions due to changes in convective motions. But are not the effects of this different for lines of different levels of formation, and therefore distinguishable from true velocity variations? 2) You have observed CO lines which are formed relatively deep in the solar atmosphere. In these layers the density is high, and by continuity the flow velocities should be relatively small. How do you reconcile your apparent velocity changes of 11 meters sec^{-1} year^{-1} with Livingston's Fe I line asymmetry changes of only 6.6 meters sec^{-1} year^{-1}?

DEMING: I do not doubt that you could distinguish between a change in convection and a center-of-mass motion; by looking at different lines you should be able to tell the difference between them. However it remains to be seen how well you can separate these two effects when they are both present.

Your comment about flow velocities only applies if there is no flow reversal in the observed region. Observationally, the highest excitation solar lines, formed deepest in the atmosphere, show the greatest convective blueshifts.

The result you refer to from Livingston applies to his data from the 1980-1982 period, which was closer to solar maximum. Any effect which is sinusoidal with the solar cycle should have a smaller rate of change at that time.

SEARCH FOR BROWN DWARFS IN THE IRAS DATA BASES

F. J. Low
Steward Observatory
University of Arizona
Tucson, Arizona 85721

Infrared observations of the self luminous planets within our own solar system (Aumann, Gillespie and Low, 1969) and the recent discovery of a substellar companion in the VB 8 system (McCarthy, Probst and Low, 1985) provide the only direct evidence for the formation of objects in the mass range 0.001 to 0.070 solar masses. There is currently no observational support for the formation of such bodies outside of larger mass systems even though their existence in large numbers is not excluded by current understanding of the processes involved. Indeed, indirect evidence (Bahcall, 1984) suggests that they may be quite numerous in the disk of our galaxy because as much as 0.1 solar mass/cubic parsec may be hidden in this form. The fact that optical searches have so far failed to detect even a single object with clear-cut substellar luminosity (Probst, 1983) and the well observed decline in the stellar luminosity function at the low luminosity end of the main sequence, suggest that if substellar objects are populous in the solar neighborhood they must be so cold that deep searches in the IR may be the only way to locate them.

The IRAS data bases provide the opportunity for systematic searches down to useful limiting magnitudes, see Table 1. Although no discoveries of brown dwarfs have been confirmed, it is of interest to include here a brief report on the initial searches and to point out some of the opportunities for future extensions in sensitivity and wavelength.

Table 2 gives the magnitudes of brown dwarfs at various temperatures and wavelengths of interest, assuming: (a) their diameters equal those of Jupiter and VB 8B, (b) they emit as simple black bodies and (c) their distances are all at 1 parsec. Table 1 summarizes the characteristics to be expected for the three main IRAS data bases and for two possible future surveys. By correlating the results in these two tables conclusions may be drawn concerning the performance of each search. For example, the IRAS Point Source Catalog cuts off at about the same temperature as the optical all sky surveys, ie 1500 K, so we do not expect to find many warm brown dwarfs in the PSC. However, a cold, very old population might be more likely to be represented given a very favorable mass function. Clearly, the most useful search requires the greater sensitivity of the coadded IRAS survey which is just now under way at the IRAS data processing center, IPAC.

Table 1. IR Searches for Brown Dwarfs

	Usable Sky Coverage (Sq. deg.)	No. of Pt. Sources	Limiting Magnitude
IRAS 12 um:			
Pt. Src. Catalog	2 E 4	>2 E 4	5
Coadded Survey	2 E 4	>4 E 4	6
Serendipity Survey	700	1500	7
Future:			
2.2 um	>2 E 4	>1 E 6	13 to 15
SIRTF 12 um	2 E 4	2 E 6	9 to 10
Dedicated Sat.	2 E 4	<2 E 7	12.5

Table 2. Brown Dwarf Magnitudes vs Temp. and Wavelength
Distance = 1 pc

T (K)	Wavelength (um)			
	0.5	0.85	2.2	12
200			38	11
300			26	9
500		37	17	7
800	38	24	11.5	6
1000	30	19	9.5	5.5
1500	20–23	13–15	7	5
2000	15–18	10–13	6	4.5
2500	12–14	8–9.5	5	4

Even though the chance of finding even a single brown dwarf in the IRAS Point Source Catalog cannot be regarded as high, it is important to make a systematic and thorough search to develop the basis for deeper searches in the future. In collaboration with colleagues at JPL, Caltech and Kitt Peak we have completed the first phase of a systematic search of the sky for absolute galactic latitudes greater than 50 degrees using the IRAS Point Source Catalog at 12 microns. Chester et al (1986) will report on the details of this work, however, the main results can be summarized as follows: (a) Virtually all of the 5776 high latitude point sources in the catalog can be positionally associated with stars or galaxies by simple comparisons with optical catalogs and atlases (b) only one object near the north galactic pole possesses a optical/IR color cool enough to be considered as a candidate, 1300 K, and it was found to be an unusual carbon star with a probable distance that places it far out in the galactic halo. Thus, interesting objects were found by selecting 12 micron sources on the basis of their optical/IR colors, classification of those sources was reasonably straight forward with only a modest amount of ground based follow up and no indications of difficulties were encountered which would impede the extension to lower latitudes and much fainter magnitudes.

A second useful part of the IRAS observations, the so called, Serendipitous Survey, is now approaching publication (Kleinmann et al, 1986). This catalog of fainter point sources is based on the coaddition of the repeated scans, each covering about one square degree, made as part of the "Additional Observations" program. Although only about four percent of the sky was covered in this mode, it provides a glimpse of the sky about 5 times fainter than the large scale survey that has been published. In order to test the quality of a preliminary version of the extraction and confirmation procedures we optically identified the 12 micron sources at high galactic latitudes.

Table 3. Preliminary Serendipity Survey Results

b > 50 ; 65 Fields ; 149 Sources ; Limiting Flux Density = 70 mJY

ID	N
Stars	119
Galaxies	19
H II Regions	6
Glob. Cluster	1
"Blank Field"	4

AG 1281	77 mJy	Star	>20 mag
AG 2157	74 mJy	Star	2' W
AG 2266	707 mJy	Star	2' E
DF 136	85 mJy	bad processing	

Those results were reported by Cutri et al (1986) and are summarized in Table 3. Once again no brown dwarfs were found, but it is interesting that the number of ordinary stars were fewer than predicted by extrapolation from the less sensitive survey and that, as might be expected, the number of extragalactic sources relative to the total increased significantly. This implies that as the IR search is extended to even lower flux levels, the predominant type of background will be individual galaxies, with their very cool and easily recognized spectral energy distributions.

Table 1 summarizes the properties of the three IRAS based efforts now under way and compares them with future undertakings: a ground based survey at 2.2 microns using very large background limited array detectors, a next generation space based survey using the Space Infrared Telescope Facility, SIRTF, and a dedicated survey satellite. Again reference to Tables 1 and 2 provides a means of judging the performance of each program in terms of limiting distance and minimum temperature.

Once the IRAS survey data have been coadded to make optimum use of the eight to twelve fold redundancy which was built in to the survey strategy, the search for brown dwarfs will enter the stage where negative results will make important statements about the formation of such bodies and about their contribution to the mass of the galaxy. Even then we may find that the search must continue, either to find more objects to study, to satisfy ourselves that such objects never form or to show that they formed so long ago that they are now extremely cold. In any event, it should be noted that the "ultimate" tool for carrying forward this quest has not yet been studied in much detail. However, it might consist of a dedicated satellite, very much like IRAS, equipped with a full compliment of background and diffraction limited array detectors (some 16 thousand pixels in each wavelength), a large on board data processor and a fully optimized survey strategy. This would extend the limits a factor of ten beyond those of the SIRTF based survey so that a 200 K object would be detected at a distance of 2 parsec.

ACKNOWLEDGMENTS

This work was supported by the National Aeronautics and Space Administration and represents the combined efforts of many of the author's colleagues on the IRAS Science Team.

REFERENCES

Aumann, H. H., Gillespie, Jr., C. M., and Low, F. J. (1969). Ap. J. Letters, 157, L69.

Bahcall, J. N. (1984).   Ap. J., <u>276</u>, 169.
Chester, et al. (1986).   In preparation.
Cutri, R., Low, F. J., Young, E. T., Kleinmann, S. G., and Gillett,
        F. C. (1985).   Bull. AAS, <u>17</u>, 878.
Kleinmann, S. G., Young, E. T., Cutri, R., Gillett, F. C., Low, F. J.
        (1986).   In preparation.
McCarthy, D. W., Probst, R. G., and Low, F. J. (1985).   Ap. J.
        Letters, <u>290</u>, L9.
Probst, R. G. (1983).   Ap. J., <u>274</u>, 237.

AN UNSUCCESSFUL SEARCH FOR BROWN DWARF COMPANIONS TO WHITE DWARF STARS

Harry L. Shipman
Physics Department, University of Delaware, Newark, Delaware
19716, U.S.A.

Abstract. A search of the IRAS database for brown dwarf companions to white dwarf stars was unsuccessful.

INTRODUCTION

The existence of low-mass, substellar objects, which we now have agreed to call brown dwarfs (Tarter 1985), in the solar neighborhood has been suspected for years, primarily because of the "missing mass" which arises from a discrepancy between the dynamically determined mass density and the mass observed in stars and other objects (Bahcall 1984, 1985). Various ways of searching for these objects have been tried. One approach to delineating the lower end of the main sequence has been to look for infrared excesses in white dwarf stars (Probst and O'Connell 1982, Probst 1983 a,b). The main idea is that a low mass companion to a white dwarf star would show up as an excess amount of infrared emission, since white dwarfs have $T(eff) > 4,000$ K.

The principal advantage of searching for brown dwarf companions to nearby objects is that the distance of the object is known from other considerations. One does not face the problem of uncovering the brown dwarf needle in a huge haystack of distant, highly reddened M supergiants or carbon stars. Furthermore, if one were to find a brown dwarf, the interpretation of the data in terms of models of these objects is considerably easier if one knows the distance directly. A disadvantage is that number of nearby objects is not that large and that if brown dwarfs are comparatively rare objects, examination of a few dozen nearby stars will fail to uncover any.

Of course, one of the stimuli for this search was the discovery of a brown dwarf companion to VB 8 (McCarthy, Probst, and Low 1985). White dwarf stars have 1/10 times the radius of late M stars like VB 8A, and consequently they will emit negligible flux in the IRAS 12 micron band. To be specific, model atmosphere calculations indicate that for temperatures ranging from 4,000 to 16,000 K, the flux in the 12 micron IRAS band for a white dwarf star only 1 parsec away should range from 0.01 to 0.08 Janskys (Jy; 1 Jy = 10^{-26} watt m^{-2} Hz^{-1}). Even a circumstellar shell, such as the one present around Vega (Aumann et al. 1984), should produce a flux of the order of only 0.01 Jy at this distance. Expected fluxes from brown dwarf companions are considerably higher, as is shown below. Thus, the brown dwarf signature in this

search is reasonably well-defined. If there is something near a white
dwarf star in an infrared all-sky survey, and it's too dim to be an M
dwarf, it is a brown dwarf.

PROCEDURE AND RESULTS

The Infrared Astronomical Satellite (IRAS), a joint project
of the United States, the Netherlands, and the United Kingdom, was
designed to conduct an unbiased survey of the sky in four wavelength
bands centered at 12, 25, 60, and 100 microns (Neugebauer et al.
1984). An array of 62 detectors in the focal plane scanned the sky as
the telescope orbited the earth, always pointing away from the earth in
order to keep the telescope cold. Successive scans were performed at
slowly increasing ecliptic longitudes so that the same region of the
sky was measured many different times. IRAS was launched in January
1983 and ceased operations in November 1983, successfully surveying 96%
of the sky.

One approach to analyzing the IRAS data was simply to compare the
catalog of spectroscopically identified white dwarf stars (McCook and
Sion 1984) with the IRAS Point Source Catalog (see. e.g., Beichman et
al. 1985). This was done at the Infrared Processing and Analysis
Center (IPAC) at the California Institute of Technology. The Point
Source Catalog contains some 250,000 infrared sources which have been
confirmed in the sense that IRAS detected them several times. This
catalog is complete to a flux density of approximately 0.3 Jy. While a
number of objects were found which were within 1.5 arcmin of the
position of white dwarf stars, all of these could be eliminated as
brown dwarf candidates for various reasons, including incorrect spectral
shape (a brown dwarf should produce the strongest signal in the 12
micron band), the existence of a previously known main-sequence companion which was the IRAS source (Sirius A and Procyon A showed up
as expected), or because the IRAS source was too bright to be a brown
dwarf, even were it associated with the white dwarf.

However, the sensitivity of IRAS can be considerably improved by
coadding the survey data. Here, several IRAS scans of a small part of
the sky are combined, point by point, in order to detect or set upper
limits to the intensity of fainter sources. Different limiting flux
densities are achieved in different parts of the sky. Coadds were
obtained for a about 20 nearby white dwarf stars, with the intent of
excluding thoses cases where a main sequence star would produce a
signal (e.g., Sirius B). The only case where a source was found at
a white dwarf position was for Stein 2051B (= EG 180, WD0426+588)
which was inadvertently included in the coadd list. Stein 2051A is an
M dwarf star which is undoubtedly the IRAS source.

In order to relate the negative results from this survey to the general
topic of brown dwarfs, consider that flux from a brown dwarf of radius
R at a distance D with temperature T can be written as

$$f \text{ (Jy)} = 0.024 \ (R/R_J)^2 \ (5 \text{ pc}/D)^2 \ (B(T)/B(2500 \text{ K})). \qquad (1)$$

In equation (1) R_J is the radius of Jupiter and B is the Planck function. The brightest reasonable brown dwarfs which could coexist in binary systems with white dwarfs, where the age of the system is constrained to be a few billion years, would have $R \sim R_J$ and a temperature near 2500 K (see, e.g., Tarter 1974, 1975; Staller and de Jong 1981, Van der Linden and Staller 1983, Nelson, Rappaport, and Joss 1985, and various theoretical papers in this volume). For some transition objects, higher values of R may not be completely excluded (D'Antona 1985). Thus, for purposes of illustrating the results of this work, it is useful to use equation (1) to relate the limiting flux densities to the quantity B_D, which is defined as follows:

$$B_D = (R/R_J) \ (B(T)/B(2500 \ K))^{1/2}. \tag{2}$$

This quantity is approximately 0.7 for VB 8B, and is of order 1 for a maximal brown dwarf. Since the Planck function at 12 microns is approximately linear in T, one can approximate the quantity B_D with reasonable accuracy as

$$B_D \sim (R/R_J) \ (T/2500 \ K)^{1/2}. \tag{3}$$

Table 1 lists the seven nearest stars where the most stringent limits to the presence of a brown dwarf were obtained. The first three columns in the table identify the star according to a name, the WD number from the catalog of McCook and Sion (1984), and the EG or Gr designation (Eggen and Greenstein 1965, 1967a, b; Greenstein 1969, 1970, 1974, 1975; Greenstein et al. 1977). The fourth column is the distance in pc, and the fifth column gives the 3-sigma limit to the flux density in the IRAS 12 micron band, for a source with a flat input spectrum. The 3-sigma limit on the detection parameter B_D is given in the sixth column. In determining this limit, the flux given in column (4) was color-corrected assuming a 2500 K black body spectrum, using standard IRAS data reduction precepts (Beichman et al. (1985). Other stars for which coadds were obtained were LP 655-42 (=WD0435-089, EG 41), G 195-19 (WD0912+536, Gr 250), Grw+70° 8247 (WD1900+706, EG

Table 1
Upper Limits to Brown Dwarf Companions to White Dwarf Stars

Star	WD Number	EG/Gr	D (pc)	Flux limit	B_D limit
V Ma 2	0046+051	5	4.2	0.12	1.6
G 240-72	1748+708	372	5.8	0.054	1.5
G 99-44	0552-041	45	6.5	0.085	2.1
L 745-46A	0738-173	54	7.0	0.075	2.1
W 489	1334+039	100	7.6	0.11	2.7
G 99-47	0553+053	290	8.1	0.085	2.6
LFT 600	0839-328	62	8.8	0.08	2.7

129), LDS 678B (WD1917-076, EG 131), G 217-37 (WD0009+501, Gr 381);
G 271-115 (WD0135-052, EG 11, L 870-2), G 99-37 (WD0548-002, GR 248),
G 107-70 (WD0727+483, Gr 52), G 60-54 (WD1257+037, Gr 95), LHS 361
(WD1345+238, Gr 438), G 226-29 (WD1647+591, Gr 368), G 208-17
(WD1917+386, Gr 375), L 997-21 (WD1953-011, EG 135), and GD 356
(WD1639+537, Gr 329). Flux limits are similar to those listed in Table
1, but because these stars are more distant, limits to B_D are less
stringent.

SUMMARY
The upper limits listed in Table 1 are close to values
which could set interesting limits to the space density of brown dwarfs
in the solar neighborhood. Even a modest improvement in sensitivity
could be very helpful. If brown dwarfs are sufficiently common that
they make up a significant proportion of the "missing mass" in the
solar neighborhood, it would be reasonable to suspect that some fraction
of them would be companions to white dwarf stars. A similar search with
a more sensitive survey instrument should yield some detections, or else
set some stringent limits on the abundance of brown dwarfs in the solar
vicinity.

One strength of using a survey like the IRAS survey in the way that was
done here, in contrast to pointed observations, is that the distance of
a putative brown dwarf companion from the white dwarf is relatively
unconstrained. In this investigation any interesting object within 10
arcmin of the white dwarf position was examined. At the distance of
the nearest white dwarf in the sample (Van Maanen 2), this is a distance
of 3000 AU. If the separation of brown dwarf-white dwarf pairs is
similar to the separation of main sequence binaries, most conceivable
possible orbits of the system will be covered.

I thank the staff of the Infrared Processing and Analysis Center,
H. H. Aumann, R. Benson, T. Chester, and H. Van Horn for assistance
and encouragement. This research was supported (in part) under NASA's
IRAS Data Analysis Program and funded through the Jet Propulsion
Laboratory. Additional support came from the National Science Founda-
tion (grants AST 81-15095 and 83-43067).

REFERENCES
Aumann, H. H., Gillett, F. C., Beichman, C., de Jong, T., Houck, J.R.,
        Low, F.J., Neugebauer, G., Walker, R.G., and Wesselius,
        P. R. (1984). Astrophys. J. (Letters) 278, L23-L28.
Bahcall, J. N. (1984). Astrophys. J. 276, 169-181.
Bahcall, J. N. (1985). Paper in this volume.
Beichman, C., Neugebauer, G., Habing, H. J., Clegg, P. E., and Chester,
        T. eds. (1985). IRAS Catalogs and Atlases - Explanatory
        Supplement (Pasadena, Calif: Jet Propulsion Laboratory).
D'Antona, F. (1985). Paper in this volume.
Eggen, O.J., and Greenstein, J. L. (1965). Astrophys. J. 141, 83-108.
Eggen, O.J., and Greenstein, J. L. (1967a). Astrophys. J. 142, 925-933.
Eggen, O.J., and Greenstein, J. L. (1967b). Astrophys. J. 150, 927-942.
Greenstein, J. L. (1969). Astrophys. J. 158, 281-293.

Greenstein, J. L. (1970). Astrophys. J. (Letters) 162, L55-L59.
Greenstein, J. L. (1974). Astrophys. J. (Letters) 189, L131-L134.
Greenstein, J. L. (1975). Astrophys. J. (Letters) 196, L117-L120.
Greenstein, J. L., Oke, J. B., Richstone, D., Van Altena, W. F., and
        Steppe, H. (1977). Astrophys. J. (Letters) 218, L21-L25.
McCarthy, D. W., Jr., Probst, R. G., and Low, F. J. (1985).
Astrophys. J. (Letters) 290, L9-L13.
McCook, G., and Sion, E. M. (1984). Villanova Univ. Obs. Contribution,
        No. 4.
Nelson, L.A., Rappaport, S.A., and Joss, P.C. (1985). Nature 316, 42-44.
Neugebauer, G., Habing, H. J., Van Duinen, R., Aumann, H.H., Baud, B.,
        Beichman, C. A., Berntema, D.A., Boggess, N., Clegg, P.E.,
        de Jong, T., Emerson, J.P., Gautier, T. N., Gillett, F. C.,
        Harris, S., Hauser, M.G., Houck. J. R., Jennings, R.E.,
        Low, F. J., Marsden, P.L., Miley, G.K., Olnon, F. M.,
        Pottasch, S.R., Raimond, E., Rowan-Robinson, M., Soifer,
        B.T., Walker, R.G., Wesselius, P.R., and Young, E. (1984).
        Astrophys. J. (Letters) 278, L1-L6.
Probst, R. (1983a). Astrophys. J. Suppl. 53, 335-349.
Probst, R. (1983b). Astrophys. J. 274, 237-244.
Probst, R., and O'Connell, R. (1982). Astrophys. J. (Letters) 252, L69-
        L72.
Staller, R.F.A., and de Jong, T. (1981). Astron. Astrophys. 98, 140-148.
Tarter, J. C. (1974). Ph.D. Thesis, University of California,
        Berkeley.
Tarter, J. C. (1975). Well-known, well-circulated, but unpublished.
Van der Linden, TH. J., and Staller, R. F. A. (1983). Astron.
        Astrophys. 118, 285-288.

DISCUSSION

D'ANTONA: When looking for brown dwarf companions to white
dwarfs we must keep in mind that limitations to their existence are
imposed by the previous evolution of the white dwarf as a giant (effective destruction of the brown dwarf, or at least significant mass
accretion, or the formation of cataclysmic variables).

SHIPMAN: It is true that a brown dwarf which was reasonably close to
the primary would be affected by the primary's previous evolution.
However, a companion which was sufficiently far away (say more than
5-10 AU away) would be reasonably unaffected. Since this survey sought
companions as far as ten arc minutes from the white dwarf, objects in
much of the parameter space which I examined would be unaffected by the
very valid points which you raise.

PROBST: Kumar has conducted a search at 3.5 microns for flux excesses
from apparently single white dwarfs. How does his approach compare to
your 12 micron search in sensitivity and areal coverage?

SHIPMAN: Dr. Kumar, would you like to comment?

C.K. KUMAR: The 3.5 micron search is much more sensitive but covers
only a small radius, about 6 arc seconds, around the white dwarf.

A DEEP CCD SURVEY FOR LOW MASS STARS IN THE DISK AND HALO

Patricia Chikotas Boeshaar
Rider College, Lawrenceville, NJ 08648

J. Anthony Tyson
AT&T Bell Laboratories, Murray Hill, NJ 07974

P. Seitzer
Kitt Peak National Observatory, Tucson, AZ 85726

ABSTRACT

As part of a deep 4-meter CCD survey in three bands, originally
begun to study the faint galaxy count problem, we are investigating statistical faint star counts and colors in our galaxy.
Most of the faint galaxies are very blue, making it possible
to search for infrared excess stars in the presence of more
than 1000 galaxies per CCD field. No extreme red dwarfs fainter than an M_V of about 15.5 (greater than 0.1 $M_\odot$) were found;
however, seven extremely red halo subdwarf candidates were
detected.

OBSERVATIONS AND IMAGE PROCESSING

Two years ago, Seitzer and Tyson began a program of 4-meter prime
focus CCD observations at CTIO, with the aim to develop techniques for
imaging and photometry to the theoretical limit of that telescope for
an exposure of 7500 sec and overall 52% efficiency: 27 J mag, 26 R mag,
25 I mag. The color-magnitude plots show that the limiting magnitudes
for 3 sigma detection are 27.2 J mag, 26.4 R mag, and 25.1 I mag.
Observed limits for 3 sigma photometry are about 1 mag brighter.

Each CCD field covers approximately 12 sq. arcmin, with total exposure
time averaging 7500 sec in each of three bands. The individual exposures are kept as short as possible (500 sec) due to time-dependent
systematics (aurora causing fringing on the CCD), and the filters were
specially designed to avoid the brighter night sky emission lines and
to be nearly identical to the effective photographic J and F responses.
In order to derive a master sky frame, the telesope is moved randomly
up to 20 arcsec between exposures, insuring that the resulting collection of images have sufficient information on the true background (sky
+ fringing) for each pixel to permit a unique determination of these
systematic errors and subtraction. These preprocessed (defringed)
images are further processed to remove bad pixels and cosmic ray events,
and are corrected for instrumental response (flat-fielded). These CCD
images, typically 15 per band, are then automatically registered to
fractional pixel accuracy and averaged, producing the final images in
each of the three bands J, R, and I. The image processing technique is

described in detail by Seitzer, Tyson, and Butcher (1985).

Processed images are next put on the VAX 11/785 at Bell Laboratories and the FOCAS v.3.2 automated detector and classifier is run, creating a catalog of properties (isophotal, aperture, and total magnitudes, centroid positions, and several central moments) for each object. After calibration with several photoelectric standards for each CCD field, the three catalogs for a given field are combined into a so-called matched catalog. These matched catalogs form the databank from which statistical inferences can be drawn. Over 2000 objects are detected in each high galactic latitude CCD frame, less than 30 of these typically are classified as stars.

In addition to the direct CCD images, artificial images are made by adding in real star and galaxy images, dimmed many magnitudes, at random locations on the CCD sky frame. The number of galaxies added follows the dex .44 mag law (Tyson and Jarvis 1979), and the number of stars added, according to what hypothesis is being tested: disk plus speroid, or disk plus spheroid plus halo. FOCAS is then rerun on this simulated data, in order to determine the probability of detection and photometric errors as a function of the magnitude of the coadded stars and galaxies. Many of these simulations are performed in order to reduce the errors.

EXTREME RED DWARFS AND SUBDWARFS

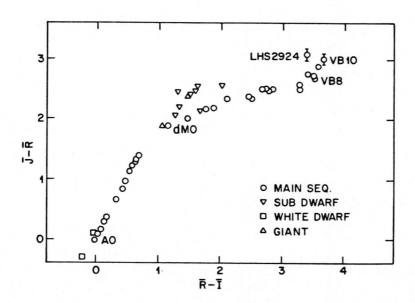

Figure 1. Colors of main sequence and M subdwarfs on the J, R, I filter system (RCA CCD).

Color data (J-R, R-I) for a sequence of calibration stars ranging from dA0 to dM9 have been obtained using the AT&T Bell Laboratories CCD camera on the 1.8 meter Perkins telescope at Lowell Observatory (Boeshaar and Tyson 1985). See Figure 1. Dwarfs earlier than M0, giants, and white dwarfs were taken primarily from the list of BVRI photoelectric standard stars by Landolt (1983). M dwarfs are from the list of spectroscopic standards by Boeshaar (1976). Extreme M sub-dwarfs (all LHS stars) have been selected from a list by Dahn, Liebert, Boeshaar, and Probst (1986) for which trigonometric parallaxes, spectrophotometry, or BVI or infrared photometry already exist. All M sub-dwarf candidates satisfy spectroscopic criteria for extremely weak metallic oxide and strong hydride features. Our broad band colors give at least as sharp a discrimination (greater than 0.4 mag in J-R at an R-I of 1.4) between the main sequence M and halo M stars as that re-ported by Liebert (1985) in B-V for a corresponding V-I.

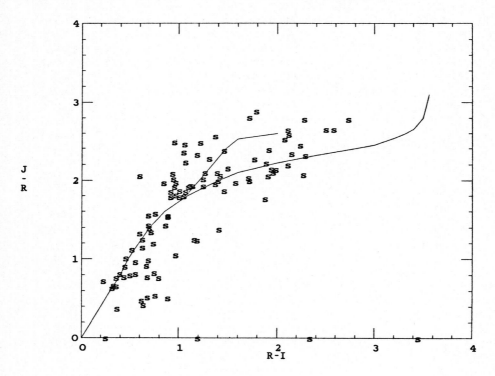

Figure 2. Objects in six CTIO CCD fields classified as stellar by FOCAS in the R and I frames, with I mags of 15 - 23. Stellar objects too faint to be detected in J are plotted along the R-I axis. The s at an R-I of 3.45 is due to a CCD defect near the edge of the frame and not an extremely red star.

We have had 7 runs on the CTIO 4-meter to date and have high quality data (average seeing FWHM ranges from 1.2 to 1.6 arcsec) on 12 high galactic latitude CCD fields. Color data on the stars found in six of these fields which had been analyzed at Bell Laboratories this past summer are presented in Figure 2. All objects plotted have been classified as stellar in the R and I bands by FOCAS. They range in I mag from 15 to 23. The mean positions of the main sequence and subdwarf sequence from the calibration data have been sketched in for comparison. The job of detecting, classifying, and photometering red dwarf and subdwarf candidates is made much safer by the absence of galaxy pollution of the portion of the J-R, R-I plane occupied by these stars. Spiral galaxies occupy the region to the lower (bluer) part of the plane, while elliptical galaxies move up (redder) and away from the stellar sequence at increasing red shift.

Note the absence of the intrinsically faintest dwarfs in Figure 2. With seven fields now completely reduced, no objects have been found with colors indicative of a red dwarf fainter than $M_V=16$. The coolest dwarfs found have I magnitudes of about 20, with color-estimated M_V's of 15-15.5 corresponding roughly to spectral type dM5-5.5. Stars of about 0.1 $M_\odot$ (i.e. UV Ceti type stars with an M_V of 16) are detectable well beyond 1 kpc at our limiting magnitudes. For detection in all three J, R, and I frames, dwarfs with an M_V of about 19 (VB10 type) can be seen out to 130 parsecs. However, assuming that an R-I greater than 3 implies one of the intrinsically faintest dwarfs, the detection limit approaches 400 parsecs. At this distance, our six fields surveyed a volume of approximately 100 pc^3. Further assuming a flat extension of the peak value of the Wielen (1974) luminosity function at $M_V=14$ to fainter absolute magnitudes, we might have expected to find 6-7 of these stars redward of R-I = 2.8. If a dynamically significant disk population, 0.5 to 1.5 times the locally observed mass density (Bahcall 1985) is in the form of stars with an M_V of 18-19, we would have expected to find 20-50 stars with an R-I = 3.4 or greater.

Some mention should be made of the scatter seen among the bluer stars. Part of this may result from the failure of the FOCAS "Splits" subroutine in splitting merged images. Since the majority of the objects in each field are blue galaxies, blended images would tend to move downward in the color-color plane to bluer J-R. As the software continues to be improved, much of the original scatter has been removed. A second source of scatter occurs in the upper half of the plot. Unresolved higher redshift elliptical galaxies are found at approximately M0 and moving vertically upward in J-R. Individual inspection of the separate JRI CCD frames for a slightly non-stellar image in the R frame is necessary to eliminate these galaxies.

Due to the unprecedented depth of this survey for stellar objects of $M_V < 18$, we have been able to sample deep into the halo for the coolest subdwarfs (M_V = 15-16 with R-I of 2, e.g. LHS 205a, LHS 1742a). At least seven such extreme M subdwarf candidates can be identified in Figure 2. If great uncertainty exist regarding the observationally

determined luminosity vs. mass relation for stars at the very bottom
of the solar abundance main sequence, the situation is even worse here.
No statement can be made regarding the masses of the M subdwarfs.
Theoretical presentations at this conference seem to suggest, at best,
that the Brown Dwarf limit is pushed to slightly higher mass in cool
stars of decreased opacity.

Only recently (Liebert 1985, Dawson 1985) have modern observational
techniques allowed the classification of these apparently ($V \geq 18$)
faint objects and a preliminary determination of their volume density,
about $5 \times 10^{-5}/pc^3$ per M_v. All extremely cool subdwarfs discovered to
date are LHS stars, i.e. discovered from photographic proper motion
surveys with a distance limit of no more than 100 pc for the intrinsically faintest subdwarfs, and 400 pc for those with an M_v of 12. Our
survey would expect to detect subdwarfs (in all three JRI bands) with
an M_v of approximately 15 out to 1 kpc. Those of $M_v=12$ are seen easily
to beyond 3 kpc. With $5 \times 10^{-5}/pc^3$ per M_v, we would expect to find
less than one subdwarf with an R-I of order 2 out to 1 kpc. Thus, the
discovery of several such candidates arouses considerable interest.
All of these candidates are 20-21 I mag, and are observable with the
4-meter spectrometers. If not subdwarfs, these red objects represent
an interesting class of object in their own right, either as red giants
beyond 50 kpc or as compact elliptical galaxies with redshifts of the
order of unity. The question naturally arises as to whether the subdwarf candidates represent a significant contribution to a massive
galactic halo. Our I band stellar number counts reveal 20 times fewer
counts than would be expected for a barely dynamically significant
halo composed of a distribution of low mass stars, i.e. R-I greater
than 1.5 corresponding to approximately $0.15 - 0.09 M_\odot$ or less.

REFERENCES

Bahcall, J. N. (1985). B. A. A. S. 17, 581.
Boeshaar, P. C. (1976). Ph. D. Thesis, The Ohio State University,
    University Microfilms.
Boeshaar, P. C. and Tyson, J. A. (1985). A. J. 90, 817.
Dahn, C. C., Liebert, J. W., Boeshaar, P. C. and Probst, R. (1986).
    (in preparation).
Dawson, P. C. (1985). B. A. A. S. 17, 560.
Landolt, A. U. (1983). A. J. 88, 853.
Liebert, J. W. (1985). B. A. A. S. 17, 545.
Seitzer. P., Tyson, J. A. and Butcher, H. (1985). A. J. (in press).
Tyson, J. A. and Jarvis, J. F. (1979). Ap. J. 230, L153.
Wielen, R. (1974), In Highlights of Astronomy, Vol. 3, G. Contopoulos,
    ed. (Reidel, Dordrecht), p. 395.

DISCUSSION

Van Horn: Can you interpret the JRI colors of your "reddest" star in terms of a temperature?

Boeshaar: The reddest of our standard dwarfs is VB10 with a T_{eff} (Reid and Gilmore 1985) of about 2600K. The reddest dwarf in our CCD data is about dM4.5 with a corresponding T_e of 3100K. For the subdwarfs, I have no idea how to scale the temperature (T_e).

Campbell: May I suggest two other possible causes of spread in your color–color plot: starspots and reddening. What is the direction of the reddening vector in that diagram?

Boeshaar: Two comments on the starspots: (1) Based on the U.S. Naval Observatory BVI photometry of a large sample of M dwarfs and subdwarfs, I would not expect to see $\pm$ 0.5 mag scatter in our color-color plot if due to star spots. Their spread in B–V is on the order of $\pm$ 0.15 mag, andprobably reflects abundance differences among the dwarfs.   (2) I seem to recall that color (B–V) fluctuations in dM0–dM1 spotted dwarfs are on the order of 0.1 mag or less.

The reddening vector runs, roughly,parallel to the dA0–dM0 main sequence.

Probst: You see no very red disk stars in your sample. A flat luminosity function predicts a few, so the null detection is marginally significant. How well constrained is a <u>rising</u> luminosity function?

Boeshaar: An extension of the Wielen Peak predicts roughly 6–7 dwarfs of M_v = 16–19 for a 100 pc^3 volume. To give you an idea of what we might expect to find for a rising log Φ :

$$\text{Wielen} - \log \Phi = -1.8 \qquad 6\text{–}7 \text{ stars}$$
$$\log \Phi = -1.65 \qquad 9 \text{ stars}$$
$$\log \Phi = -1.5 \qquad 12 \text{ stars}$$

Unless these dwarfs are all part of close binaries, we should have detected at least 1 in 12.

D'Antona: Do I understand correctly that if you take a decreasing luminosity function for $M_v > 14$, as it appears for nearby stars, you do not expect any in your samples.

Boeshaar: The number expected would not be statistically significant.

Kafatos: So, again, your limiting magnitude is 25 per 12 square arcmin and you see approximately 2000 galaxies down to that limit?

Boeshaar: No, each CCD field is 11–12 square arcmin. The limiting magnitude for photometry (0.2 mag error) for 7500 sec summed exposures in each band are: $J \sim 26$ mag, $R \sim 25$ mag, and $I \sim 24$ mag. About 1000 galaxies are photometered in each field, to the above limits. Over 2000 objects are detected to J=27 mag in one 12 sq. arcmin field.

AN INFRARED SEARCH FOR LOW-MASS COMPANIONS OF STARS WITHIN 12 PARSECS OF THE SUN

M. F. Skrutskie
Cornell University, Ithaca, N.Y. 14853

W. J. Forrest
University of Rochester, Rochester, N.Y. 14627

M. A. Shure
University of Rochester, Rochester, N.Y. 14627

Abstract. Sixty stars within 12 parsecs of the Sun were observed at $2.2\mu m$ (K band) using the Rochester 32×32 element InSb infrared array camera. The search covered a square field 14 arcseconds on a side centered on the target stars and was sensitive to a limiting magnitude (5σ) of 15.0 at K. This magnitude corresponds to a dwarf companion with an effective temperature of 1000K at 10 parsecs, well below the effective temperature of about 2000K at the low-mass hydrogen burning limit for main sequence stars. Owing to the saturation of the primary star image, the inner radius of the area searched was 2 arcseconds from the primary. Deconvolution techniques may eventually recover some sensitivity inside this radius. No substellar companions were detected around any of the stars, indicating a lack of companions down to roughly $0.04M_\odot$. This result implies that the main sequence luminosity function, at least for close companions, probably turns downward below M_V of 16, and indicates that the dark disk mass must reside in forms other than substellar companion objects in the mass range above. In addition, eight stars in a young galactic cluster, the Pleiades, were imaged to a limiting magnitude of 15.7 with the same result. The relative youth of this cluster ($<10^8$ years) implies that substellar objects should still be very warm following their initial contraction. The overall lack of detections suggests that substellar brown dwarfs may be quite rare.

INTRODUCTION

Interest in very low mass stars and substellar objects stems from their potentially important role in several longstanding astronomical problems, especially as a possible source of dark matter in galaxies. In particular, our limited knowledge of the shape of the faint end of the stellar luminosity function and its extrapolation to substellar masses leaves open the possibility that low mass objects may account for the local Oort mass (Bahcall 1984). To date, deep optical searches (Boesharr & Tyson 1985) for low mass field objects have indicated that low mass stars are indeed rare and that the stellar luminosity function probably does decrease for $M_V > 16$. Constraints on substellar objects, which have lower effective temperatures and luminosities, are not so strong.

The predominance of binary stellar systems suggests another approach to the problem, namely searching for faint companions in the vicinity of nearby stars. If low mass companions constitute the dark disk mass then they should be quite common. A search of stars in the neighborhood of the Sun would be quite sensitive, especially if the wavelength chosen coincided with the blackbody emission peak of these cool objects (i.e. the near infrared). Infrared speckle interferometry (McCarthy 1982) allows searches for companions on small spatial scales, but sacrifices several magnitudes of sensitivity in order to do so. Since the distribution of binary separations favors spacings of a few arcseconds for companions to nearby stars, direct imaging with infrared detectors may be more productive. Jameson *et al.* (1983) carried out such a search on 21 stars with a single element detector and found no companions.

Infrared arrays which are capable of imaging large fields of view now exist, considerably increasing the sensitivity of such a search. This paper reports preliminary results for a survey of 60 stars within 12 parsecs of the Sun as well as 8 stars in a young nearby galactic cluster (the Pleiades) using the Rochester InSb CCD Array Camera in the K photometric band (2.2μm). For the nearby stars, the limiting (5σ) sensitivity corresponded to an effective temperature of 1000K, well into the substellar regime. Nevertheless, the survey uncovered no previously unknown companions around any of the objects searched. Either substellar companions are quite rare, or else they cool so rapidly as to be undetectable in this survey.

OBSERVATIONS

Observations were made at the 3m IRTF telescope on Mauna Kea on 26-28 July 1985 with the University of Rochester InSb CCD Array Camera (Forrest *et al.* 1984). The 32×32 element square array subtended a field of view 14" on a side (.47"/pixel). All stars were centered on the array and imaged through a standard K band photometric filter with a total integration time of 5 minutes (eight 40 second frames averaged together), corresponding to a limiting (5σ) K magnitude of 15.0. If the K band image showed a potential companion additional frames were taken at H (1.6μm) in order to determine the color temperature of the source. A blank field (typically 100" south of the star) served as a reference for background subtraction. A pair of 40 second images on the background field followed each pair of integrations on the source.

Saturation of the primary star image and the wings of the gaussian seeing profile limit the innermost radius searched to about 2". In addition, the brightest primaries produced scattered light in the instrument and telescope, limiting the sensitivity away from the star. This effect was most pronounced in the southwest corner of each frame. Scattered light was not a problem for primaries of spectral type M3 and later. The survey emphasized this range of spectral types, and avoided primaries with $K<6$ as much as possible. Despite this bias, the sample reflects the distribution of spectral types in the solar neighborhood, owing to the predominance of late type stars in the luminosity function.

The success of such a search depends strongly on the seeing, in terms of both sensitivity and the spread of the primary image. Ultimately, all of the light from the star should fall on a few pixels. Seeing was about one arcsecond for the majority of the targets. The seeing was quite stable, allowing the use of one image as a point spread function for the deconvolution of another image. Preliminary attempts at CLEANing and Maximum Entropy techniques have provided promising results. Future analysis should recover some sensitivity in the innermost two arcseconds, as well as slightly improve the signal-to-noise over the rest of the map.

The survey consisted of 60 stars chosen from the Wooley (1970) catalog of stars within 25 parsecs of the Sun. In order to be a candidate the star must have been within 12 parsecs of the Sun and at least 20° from the galactic plane. Confusion from background sources was minimal. In addition, the survey contained a sample of 8 stars from the

Pleiades cluster. The relative youth of this cluster ($< 10^8$ years) permits the detection of substellar objects which are still quite warm following their initial contraction. The greater distance of this cluster (120 parsecs) extends the search radius to the range of 200 to 1000AU from the primary. Deeper integrations partially compensated for the greater distance.

The 15.0 magnitude detection limit (5σ) at K corresponds to a Jupiter-sized blackbody with an effective temperature of 1000K at a distance of 10 pc. Degenerate cooling models (Nelson *et al.* 1985) at an age of 5×10^9 years imply a limiting mass of $0.04M_\odot$. At the Pleiades, the 15.7 magnitude limit corresponds to an effective temperature of 2200K. Although this temperature is much greater than the survey limit at 10 parsecs, the youth of this cluster compensates to make the search in the Pleiades equally sensitive to substellar objects.

The results of the survey are easily summarized: No cool companions were detected around any of the stars within the limits cited above.

DISCUSSION

The lack of detection of any substellar companions in this survey is quite striking, especially since the binary star semi-major axis distribution peaks in the survey range of 20 to 100AU (Heintz 1969; Abt & Levy 1976). An examination of the probability that a given primary will have a substellar companion detectable in the survey will clarify this result. The probability that a star will have a companion is complicated by the fact that the above semi-major axis range spans the transiton between bifurcation binaries and capture binaries (Abt 1979). The capture binaries, forming separately, are likely to follow a distribution similar to the field star luminosity function. Bifurcation doubles tend to favor companions of equal mass (Lucy & Ricco 1979). Since the greatest amount of areal coverage in the survey coincides with the capture binary domain, we will analyze the results mainly in terms of constraining the field star luminosity function.

Given a primary of mass, m_p, the probability that it will have a substellar companion with mass, $0.08M_\odot > m_c > 0.04M_\odot$, and lie at a radius between 20AU and 100 AU is:

$$P = \frac{\int_{0.04}^{0.08} \phi(m)\, dm \int_{20AU}^{100AU} \psi(r)\, dr}{\int_0^{m_p} \phi(m)\, dm \int_0^\infty \psi(r)\, dr} \eta(m_p)$$

where ϕ is the mass function, ψ is the semi-major axis distribution, and η is the probability a star has a companion as a function of its mass. The mass function, ϕ, is the least well known of the terms. The degree to which the other terms are understood will constrain the extrapolation of the mass function to substellar objects. The above equation applies if all the terms are independent. The limited statistics presently available for binary stars suggest that this is a reasonable assumption.

The probability, η, that a given star has a binary companion is roughly independent of spectral class (Heintz 1969), although there is some evidence for a decrease in binary frequency for lower mass objects. Since the mass of the primary defines the upper limit for the companion mass, this trend may be evidence for a greater fraction of dark companions around late-type stars. On the other hand, low mass stars may be inefficient at capturing companions and the deficit may be a real effect. In any case, a value of $\eta=0.4$ is consistent with the observations.

The semi-major axis distributions function, ψ, is fairly well determined (Heintz 1969; Abt & Levy 1976), and seems to be independent of spectral class. Roughly 25% of binaries will have projected separations between 20 and 100AU.

The mass functions, ϕ, is very poorly determined in the mass range of interest. Current evidence for field stars (Miller & Scalo 1979) indicates that the mass function is constant, or possibly decreases below $0.1M_\odot$. If the mass function is flat, the fraction of stars with companions which lie between $0.04M_\odot<m_c<0.08M_\odot$ is about 25%, assuming a lower mass cutoff of $0.02M_\odot$ and a primary mass of $\sim0.3M_\odot$. Combining this result with the other terms predicts 2 detections in the survey, marginally ruling out a flat mass function for companions below $0.1M_\odot$.

More importantly, the amount of dark matter in the galactic disk is of the same order the mass seen in stars and interstellar matter (Bahcall 1984). If substellar companions account for the dark disk mass, then every star should have a roughly equal mass in dark companions. In this case, the probability of detection is mainly determined by the semi-major axis distribution, and the survey should have found 10 to 15 candidates. The null results rule out the possibility that dark companions harbor the dark matter, unless the majority of them have masses less than $0.04M_\odot$.

CONCLUSION

Near infrared imaging of 60 nearby stars and 8 stars in the young Pleiades cluster uncovered no substellar companions down to a limit corresponding to a mass of $\sim0.04M_\odot$. The dark matter in the galactic disk cannot reside solely in substellar companions unless it is largely in objects less massive than the survey limit. The results marginally exclude a flat mass function for companions below $0.1M_\odot$, consistent with the suggested maximum in the field star mass function.

This survey would greatly benefit from extended spatial coverage around the primary stars and sensitivity to lower effective temperatures. Infrared imaging technology is rapidly progressing and larger arrays with better sensitivity promise to be available in a short time.

REFERENCES

Abt, H.A. (1979). *Astron. J.*, **84**, 1591.
Abt, H.A. & Levy, S.G. (1976). *Ap. J. Supp.*, **30**, 273.
Bahcall, J.N. (1984). *Ap. J.*, **276**, 169.
Boesharr, P.C. & Tyson, T. (1985). *Astron. J.*, **90**, 817.
Forrest, W.J., Moneti, A., Woodward, C.E., Pipher, J.L. & Hoffman, A. (1985). *Pub. Astron. Soc. Pacific*, **97**, 183.
Heintz, W.D. (1969). *Journal Royal Astr. Soc. Canada*, **63**, 275.
Jameson, R.F., Sherrington, M.R. & Giles, A.B. (1983). *M.N.R.A.S.*, **205**, 39P.
Lucy, L.B. & Ricco, E. (1979). *Astron. J.*, **84**, 401.
McCarthy, D.W., Low, F.J., Kleinmann, S.G. & Gillett, F.C. (1982). *Ap. J. Letters*, **257**, L7.
Miller, G.E. & Scalo, J.M. (1979). *Ap. J. Supp.*, **41**, 513.

Nelson, L.A., Rappaport, S.A. & Joss, P.C. (1985). *Nature*, at press.
Wooley, R., Epps, E.A., Penston, M.J. & Pocock, S.B. (1970). *Royal Obs. Annals*,
      No. 5.

REMARKS

<u>Shipman</u> : The discovery of additional white dwarf companions to main sequence stars would be quite interesting. While there are roughly 1200 known white dwarfs (and thus the discovery of another garden variety one would be relatively uninteresting, there are very few (only 6, specifically) in binary systems where the mass can be determined. Was that Gliese 816 which had a possible white dwarf companion?

<u>Skrutskie</u> : Yes. The companion has a luminosity consistent with a 10,000K white dwarf at the distance of Gliese 816. The separation between the two components is about 7" or 85AU. The period is then of order a few hundred years, so it may be some time before an accurate mass is available.

<u>Low</u> : Can you subtract away the emission from a point source to improve your chance of finding companions within 2 arcseconds?

<u>Skrutskie</u> : Since the seeing remained stable throughout most of the observations I believe we have a good chance of doing so. I have used a CLEAN algorithm on some of the maps and the results have been encouraging.

THE BERKELEY SEARCH FOR A
FAINT STELLAR COMPANION TO THE SUN

S. Perlmutter, M. S. Burns, F. S. Crawford, P. G. Friedman,
J. T. Kare, R. A. Muller, C. R. Pennypacker, and R. W. Williams

Lawrence Berkeley Laboratory
University of California, Berkeley, CA 94720

Abstract. Based on considerations of wide-binary-star stability, it is
reasonable to look as far as 1 to 2 pc for a stellar companion to the
sun. We have developed an automated telescope and image
analysis system for finding such a high-parallax star among many
faint candidates. We will detect (at 3σ) parallactic motion $>0.75''$,
corresponding to a companion closer than 2.6 pc. Our candidate
list consists of ~5000 late M stars, but our current system would
allow us to include any additional sources down to $m_v = 16.5$, or a
Jupiter-size black body of 1600 K at our limiting distance. We
have collected ~90% of the first epoch images, and have tested the
image processing on a hundred fields taken months apart.

INTRODUCTION

Studies of binary stars suggest that the fraction of solar-type stars
in binaries is probably greater than 84% (Abt 1983). It is usually assumed that
the sun is an exception, perhaps because it is orbited by the next best thing, a
planetary system. Occasional articles have suggested that a solar companion star
could exist and remain unseen or perhaps be seen and not identified (Reynolds,
Tarter & Walker 1980; van de Kamp 1971, 1982). Knowledge of the sun's binary
status would add to our understanding of star formation. If a companion were
found, we would also have an important example for stellar evolution studies.

Recently, it has been suggested that a companion star could explain 26-million-
year periodicities found in mass extinctions of species on the earth (Davis, Hut &
Muller 1984; Whitmire & Jackson 1984). The mechanism has since been further
elucidated and analysed (Muller 1984; Hut 1985), but the final ruling on the
hypothesis may have to await the outcome of a search for such a companion.
This need has provided the stimulus for the Berkeley search for a stellar companion to the sun.

PLAUSIBLE LIMITING DISTANCE FOR A COMPANION

How far away is it reasonable to expect to find a solar companion? Essentially this is a matter of how fast a binary pair at a given separation distance is expected to be broken up by passing stars, molecular clouds, and so on. Observationally, there is a steep drop off in number of binaries with a semi-major axis wider than 0.1 pc. Bahcall & Soneira (1981) have shown that this is not just a selection effect by analyzing pair correlations and nearest-neighbor distributions for stars brighter than $V = 16$ near the North Galactic Pole.

This observational cutoff distance is in agreement with both theoretical expectations and numerical simulations for binaries in the solar neighborhood. For conditions in the solar neighborhood, the half-life of a wide binary is found analytically to be

$$t_{1/2} \approx \frac{10^{10} \text{ yr}}{(a / 0.1 \text{ pc})}$$

where a is the semi-major axis (Bahcall, Hut & Tremaine 1985; Retterer & King 1982). Thus in the 10 billion year life of the galaxy most binaries wider than $a = 0.1$ pc would be expected to have broken up. The numerical simulations of Bahcall, Hut & Tremaine (1985) indicate comparable lifetimes.

What does this tell us about the plausible range of distances for a companion to the sun? To a first approximation, we might expect to find a companion only at distances $a \leq 0.2$ pc where the remaining lifetime is about the age of the solar system. This limit, however, does not take into account the fact that a companion star probably formed with the solar system, and thus has suffered 4.5 billion years of perturbations from its initial orbit. (The probability for a more recent capture of a passing star is very small.) These perturbations would have caused the solar binary to trace a random walk in its semi-major axis, on average becoming wider at an increasing rate. Figure 1 shows a schematic representation of this process, based on numerical simulation results of Hut (1984).

Thus if a companion star still exists it would be expected to have random walked out from this initial separation and therefore be likely to have much less than its initial lifetime left. Assuming we are not very unlucky and the companion is just about to disappear, we should look for it at distances where the remaining lifetime is still an appreciable fraction of the age of the solar system. If we choose a remaining expected lifetime of 1/10 to 1/4 the age of the solar system, the corresponding semi-major axes are smaller than about 1 to 2 pc. This gives us a ballpark figure for the maximum distance at which it is worth looking for a solar companion.

THE BERKELEY AUTOMATED SEARCH

Our astrophysics group at Lawrence Berkeley Laboratory has found itself in a particularly opportune position to conduct a search for such a companion star. We have developed an automated system to be used for a supernova search, including a computer-controlled telescope and extensive image analysis software (Kare *et al.* 1981). This gives us the capability to collect and store hundreds of digitized CCD images each night, and then, automatically to compare images of different nights in software. Such a system is exactly what we need for a parallax search for a solar companion. Instead of comparing images taken nights apart, looking for a new bright star, we compare images taken months apart, looking for parallactic motion of candidate stars compared to background stars in the same field.

Our current system has a scale of 2.5″ per CCD pixel, and we fit the center coordinates of the stellar images to an accuracy of 1/10 of a pixel. We can thus, with 3σ confidence, find parallactic motion greater than 0.75″. The limiting distance for our search is then approximately 2.6 pc for a six month search. This distance includes the companions with expectations of a significant fraction of a solar lifetime remaining. It also includes the 0.8 pc distance expected for a companion which could explain a 26 million year period for mass extinctions.

Figure 1. Schematic representation of the random walk in semimajor axis of a solar companion.

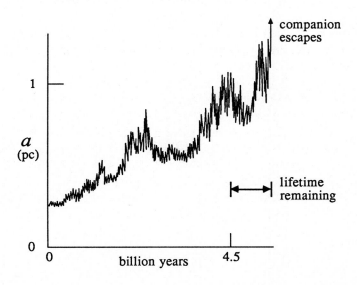

CANDIDATES FOR THE SEARCH

To use this method one needs lists of stars which are likely candidates to have their parallaxes measured. The Yale Bright Star Catalogue, complete to $m_v = 6.5$, rules out any stars brighter than this apparent magnitude, since it contains no entries with the kinematic and photometric characteristics possible for a solar companion. A companion at 1 pc must then be fainter than absolute magnitude $M_v = 11.5$, and we are therefore only interested in the lower part of the H-R diagram containing late M dwarfs, some cooler white dwarfs, and brown dwarfs.

Our search is currently covering the late M stars in the northern hemisphere. We are using the Dearborn Observatory Catalogue of Faint Red Stars, which is apparently complete for late M stars with about 5000 catalogued between M4 and M8. We are not currently including any white dwarfs in our search, partly because there is no comparable survey for white dwarfs. Current models of white dwarf evolution suggest that in the lifetime of the solar system a white dwarf companion would have cooled to an apparent magnitude between the $m_v = 6.5$ cutoff set by the Yale Bright Star Catalogue and the $m_v = 16.5$ detection limit of our current system (Shapiro & Teukolsky 1983). Any cool white dwarfs ($< 10,000$ K) lacking a parallax measurement would therefore be worth adding to our list, although presumably these stars are interesting enough for other reasons that they will be examined thoroughly one at a time.

We are of course interested in including brown dwarf candidates. Our current detector configuration will allow us to see a Jupiter-size blackbody as cool as 1350 K at 1 pc, or 1600 K at our limiting distance. If we use the cooling rates presented elsewhere in these conference proceedings, it seems unlikely that a 4.5 billion-year-old companion would be visible with our system.

CURRENT STATUS

We have collected about 4500 first epoch images of stars from the Dearborn Observatory Catalogue and have run the processing software on some 100 sample fields taken on different months. In the meantime, we have been working on a number of improvements on the system: First, the automated telescope speed was increased to allow observation of 5000 images in less than two weeks. Second, we are testing a new CCD from RCA with promised better performance. Third, we are changing the scale to $1''$ per pixel, making it is easier to obtain the accuracy we want. To take advantage of these modifications, we will collect many of the first epoch images over again.

REFERENCES

Abt, H. A. (1983). Normal and abnormal binary frequencies. *Ann. Rev. Astron. Astrophys.*, **21**, 343-72.

Bahcall, J. N. and Soneira, R. M. (1981). The distribution of stars to $V = 16$th magnitude near the North Galactic Pole. *Astrophys. J.*, **246**, 122-135.

Bahcall, J. N., Hut, P., and Tremaine, S. (1985). Maximum mass of objects that constitute unseen disk material. *Astrophys. J.*, **290**, 15-20.

Davis, M., Hut, P., and Muller, R. A. (1984). Extinction of species by periodic comet showers. *Nature*, **308**, 715-717.

Hut, P. (1984). How stable is an astronomical clock which can trigger mass extinctions on the earth? *Nature*, **311**, 638-641.

Hut, P. (1985). Evolution of the solar system in the presence of a solar companion star. In *The Galaxy and the Solar System*. Proceedings of conference, Tucson, Arizona, January 1985, to be published.

Kamp, van de, P. (1971). The nearby stars. *Ann. Rev. Astron. Astrophys.*, **9**, 103-126.

Kamp, van de, P. (1982). Evolutionary trends in wide binaries. In *Binary and Multiple Stars as Tracers of Stellar Evolution*, ed. Z. Kopal and J. Rahe, pp. 81-103. Dordrecht: Reidel.

Kare, J. T., Pennypacker, C. R., Muller, R. A., Mast, T. S., Crawford, F. S., and Burns, M. S. (1981). The Berkeley automated supernova search. In *Supernovae: A survey of current research*, ed. M. J. Rees and R. J. Stoneham, pp. 325-339. Dordrecht: Reidel.

Muller, R. A. (1984). Evidence for a solar companion star. *Lawrence Berkeley Lab. Report* LBL-18271. Also in Proceedings of the Int. Astron. U. Symposium 112, to be published.

Raup, D. M. and Sepkoski, J. J. (1984). *Proc. Nat. Acad. Sci. U.S.A.*, **81**, 801-805.

Retterer, J. M. and King, I. R. (1982). Wide binaries in the solar neighborhood. *Astrophys. J.*, **254**, 214-220.

Reynolds, R. R., Tarter, J., and Walker, R. G. (1980). *Icarus*, **44**, 772-779.

Shapiro, S. L. and Teukolsky, S. A. (1983). *Black Holes, White Dwarfs, and Neutron Stars*. New York: Wiley.

Whitmire, D. P. and Jackson, A. A. (1984). Are periodic mass extinctions driven by a distant solar companion? *Nature*, **308**, 713-715.

DISCUSSION

Liebert: Wouldn't it be rather inexpensive in time to do filter CCD photometry at one of your epochs on each star (or separately with a photoelectric photometer) and catalogue the B, V and I magnitudes? First, there is clear differentiation between giants and dwarfs from such a two color diagram. Second, the breakdown of such a "complete" sample including oxygen and carbon giants would be useful.

Bahcall: Your survey would be of great value to galactic structure and content studies if you made a catalogue of all the objects with photometric magnitudes (accuracy <0.05 magnitude), and possibly colors. Is that something you could do without greatly increasing your effort?

Perlmutter: It is currently easier for us to distinguish the dwarfs from the giants by parallax than by color. We are however installing an automated filter wheel, and will study the uses of a multicolor photometric catalogue of our stars. Since most of our observing time is spent integrating light rather than pointing, the photometric data can be collected separately without loss of time.

Roman: I am under the impression that many stars in the Dearborn catalogue are carbon stars. 1) Have you looked for spectroscopic data for your stars? 2) What is the magnitude limit of the catalog? 3) How does this source compare with the data from Schmidt surveys for M dwarfs?

Perlmutter: The Dearborn catalogue was made from an objective prism survey on red sensitive plates, and gives spectral types for its stars. We are only searching the stars which are classified as M4 through M8. The catalogue claims to be complete down to 13th magnitude, and this seems to be confirmed by more recent red dwarf surveys, such as that of C. B. Stephenson at Case Western.

Lindsay: It has been pointed out that evidence for periodicity in the fossil sample comes from binning the data in 6.2 m.y. bins and identifying an episode of extinction as a bin with more extinctions than the preceding and following bins—a technique which will necessarily lead to a peak in the power spectrum at about four times the bin width. Are you not discouraged by this observation?

Perlmutter: Raup & Sepkoski (1984) answered this criticism in their original paper on extinction periodicity. Their Monte Carlo calculations showed that the significance of the extinction data periodicity was greater than that of all of their 8500 randomized data sets *binned the same way,* giving a confidence level greater than 99.9%. Muller (1984) has shown that an alternative analysis of the same data that avoids binning the data gives essentially the same result.

THE LUMINOSITY FUNCTION FOR LATE M-STARS

M.R.S. Hawkins
Royal Observatory, Blackford Hill, Edinburgh, Scotland

(Present address: UK Schmidt Telescope, Private Bag,
Coonabarabran, NSW 2357, Australia)

Abstract. A search for low luminosity M-stars is
described. A sample is obtained on the basis of (R - I)
colour which gives a space density distribution similar to
earlier work for $M_R < 14$, but shows a subsequent increase
in space density for the faintest stars. The implications
of these stars for the problem of missing mass in the
neighbourhood of the sun is discussed. It is concluded
that these stars form part of a population of substellar
brown dwarfs which may make up the mass deficit in the
solar neighbourhood.

In this talk I want to describe the rather circuitous route which has
led, I believe, to the identification of the local missing mass. Since
Oort (1965) first pointed out the discrepancy between directly observed
mass and that deduced from stellar kinematics, a wide variety of
explanations have been suggested, many of which are of a somewhat
speculative nature. It is perhaps surprising that the most obvious
candidates, namely faint low mass stars, were rejected so readily. The
benchmark for studies of the faint end of the stellar luminosity
function has been the distribution from Luyten's (1968) proper motion
surveys, and appears to show a cut-off at a luminosity of about $M_V =$
14-15, although as Luyten is at pains to point out, the numbers are
very uncertain in this region. Sanduleak (1976) has challenged this
finding with the claim that there is a substantial number of low
velocity M-dwarfs missed by the proper motion searches, but there seems
general consensus among other astronomers that the luminosity function
does turn over as indicated by Luyten. This conclusion is based on
several studies, including that of nearby stars by Wielen (1974), and
more recently the photometric survey of Reid & Gilmore (1982), who make
the further claim that their observations rule out missing mass in the
form of faint stars, at least to an apparent magnitude of I = 17.

The present work was originally planned as a proper motion search for
faint white dwarfs. The search was based on sets of deep UK 1.2m
Schmidt Telescope plates, of the ESO/SERC field 287 at $21^h 28^m$, -45°
(1950), measured by the COSMOS automatic measuring machine at the Royal
Observatory, Edinburgh. This machine has the capability of scanning a
35cm plate in a few hours, detecting images above a specified threshold
isophote and outputting 20 parameters for each image including position
and magnitude measures. Some 150,000 images were detected in 18 square

degrees, to a completeness R = 19.8 on IIIa—F emulsion through an RG630 filter. The measures were calibrated photometrically as described by Hawkins (1984) and references therein, using a combination of photoelectric, electronographic and CCD standards, and some 50 plates in UBRI were used. The resulting photometric accuracy was about 0.1^m per plate, giving internal errors for each star of 0.05^m or better in all colours. The positional measures after transformation to a common coordinate system gave residuals of about 2.5 microns per plate and, since several plates were available in two epochs, displacements were obtained accurate to 1.5 microns or 0.1 arcseconds.

In the first instance, blue passband plates were used in the search for white dwarfs, and the results turned out much as expected, both for the white dwarfs and also for main sequence stars. In order to extend the search to redder objects the analysis was repeated on the red passband plates, and in this case the results were not as expected. The overwhelming majority of the sample was of course composed of M—dwarfs, but among these were some 10 stars with apparently extremely red colours [(R — I) > 2] and 3 with (R — I) > 2.7. In fact some of these objects had turned up in other unrelated searches, and been dismissed as very red compact galaxies. The proper motion measures made it clear that this classification was incorrect, and suggested a hitherto undetected population of very low luminosity stars. Furthermore, the close alignment of the proper motion vectors along the reflex of the solar motion seemed to imply a very small velocity dispersion, and enabled an estimate of distance on the basis that the stars were essentially showing the solar motion. This distance agreed well with that from the absolute magnitude obtained from the (R — I) colour. The only rather worrying thing was that the space density obtained in this way was embarrassingly high, in itself enough to make up the local missing mass.

This result was treated with some scepticism by several astronomers working in the same area, both with regard to the rather rough kinematic distance measure, and to whether the colours really were as red as claimed. As it turned out, both those doubts were justified. The first attempt by Dr M S Bessell to check the colours photoelectrically immediately indicated that something was wrong. After considerable investigation it emerged that the colour equation from the IIIa—F/RG630 passband to the Cousins system was not zero as stated by Reid & Gilmore (1981) and Blair & Gilmore (1982), but very substantial indeed, up to 0.6 magnitudes for the reddest stars. Although it seemed unlikely that the large space density of low luminosity stars would actually disappear with a recalibration of the R magnitudes, it was apparent that to provide an unambiguous picture of variations in space density it was necessary to measure the luminosity functions for the whole M-star population.

It was decided to follow the photometric approach advocated by Reid & Gilmore (1982), but with some important additions and improvements. Firstly, the sample would be selected on the basis of (R — I) colour rather than (V — I), allowing an improvement in limiting magnitude of 1.5^m. Only stars detected in both colours would be counted, obviating

the need for conclusions based on non—detections. Also, the
photometric data would be supplemented by proper motions enabling the
measurement of tangential velocity, and velocity dispersion. Wherever
possible the data of Reid & Gilmore would be used to facilitate
comparison between the two samples.

The procedure for selecting the new sample was as follows. First, all
objects with $(R_F -I) > 0.8$ were selected, the F suffix indicating the
IIIa-F/RG630 passband. The $(R - I)$ versus M_R relation was taken from
the photometry of parallax stars by Reid (1982) and the conversion from
the Cousins R to R_F magnitudes was carried out using the relation
derived by Bessell (1985). These results when applied to the $(R_F - I)$
and R_F measures of the sample gave absolute magnitude M_{RF}, and distance
for each star; following Reid & Gilmore (1982), stars outside 100 pc
were rejected. The sample was complete to 100 pc down to $M_{RF} = 15$, and
the faintest bin of $M_{RF} = 15 - 16$ was complete to 60 pc. In order to
facilitate comparison with Gilmore and Reid's work, R_F magnitudes were
converted to V, again using the results of Reid (1982) and Bessell
(1985). Figure 1(a) shows a histogram of absolute magnitude M_{RF} for
the sample, and 1(b) a similar histogram for the corresponding V
magnitudes. The figures show marked minima at $M_{RF} = 13$ and $M_V = 15$.
The shaded portion in the M_V histogram shows stars with $V > 19.6$, the
completeness limit of Reid & Gilmore's (1982) V plates. Figure 2(a)
shows a plot of space density versus M_{RF}. The filled circles show the
sample with $\sqrt{N}$ error bars, and the open circles are the same data but
rebinned at a difference of half a magnitude. Figure 2(b) shows space
density as a function of M_V, where the V magnitudes were obtained as
described above and are shown as filled circles. The crosses show the
data of Reid & Gilmore (1982) and it will be seen that from $M_V = 12 -$
16 the agreement between the two samples is remarkably good. For
brighter magnitudes there is an excess of objects in the present
sample, which is attributable to the difference in galactic latitude
between the two fields causing differential contamination by halo
giants. Reid & Gilmore (1982) only have two stars fainter than $M_V = 16$
in their sample, but perhaps the most remarkable thing is that taken at
face value they appear to show the same trend as the present sample.
Before discussing the significance of Figure 2(a) we first examine the
proper motion data.

The availability of some 15 deep IIIa-F plates in two epochs separated
by 4 years made possible the measurement of proper motion for every
star in the sample. These measures combined with the photometric
distances gave tangential velocities which are shown in Figure 3 as a
function of M_{RF}. The error on the velocities is essentially due to the
error in the proper motions and is about 7 km/sec for most of the
sample, reducing to 4 km/sec for the faintest stars which are somewhat
nearer.

There is no indication in Figure 3 of a significant trend of velocity
with magnitude, although the five bright high velocity stars tend to
distort the picture slightly. Figure 4(a) shows velocities for the
sample plotted in the plane of the plates. The direction of galactic
rotation (+v) is almost directly due north (+y), while the +x direction

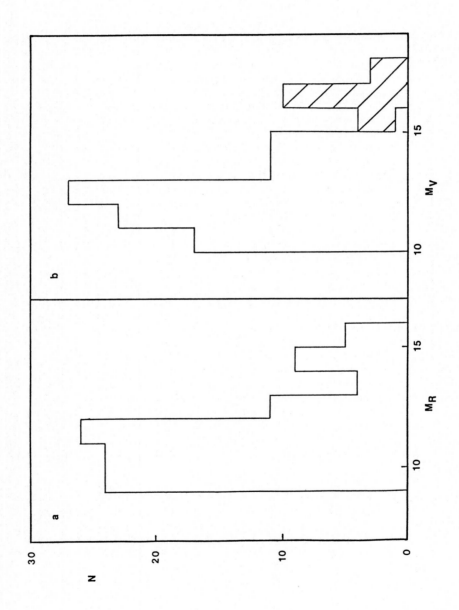

1. Histograms of the sample of M–stars for (a) absolute R magnitude (M_{RF}) and (b) absolute V magnitude (M_V). The shaded areas are stars with V > 19.6.

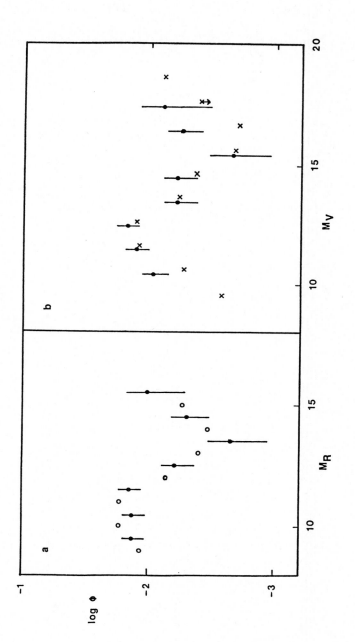

2. Plots of log space density in stars per pc^3 (a) for absolute R
    magnitude (M_{RF}) shown by filled circles with $\sqrt{N}$ error bars. The
    open circles are the same data rebinned by half a magnitude; (b)
    for absolute V magnitude (M_V). The filled circles are for the
    present sample, and crosses from Reid & Gilmore (1982).

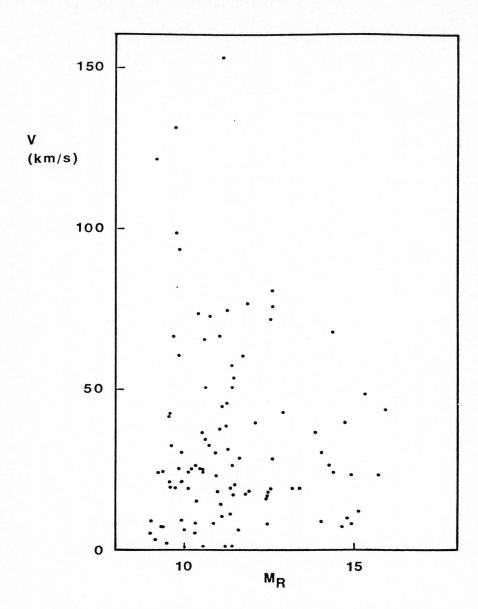

3.   Plot of velocity versus absolute R magnitude (M_{RF}) for the whole
sample.

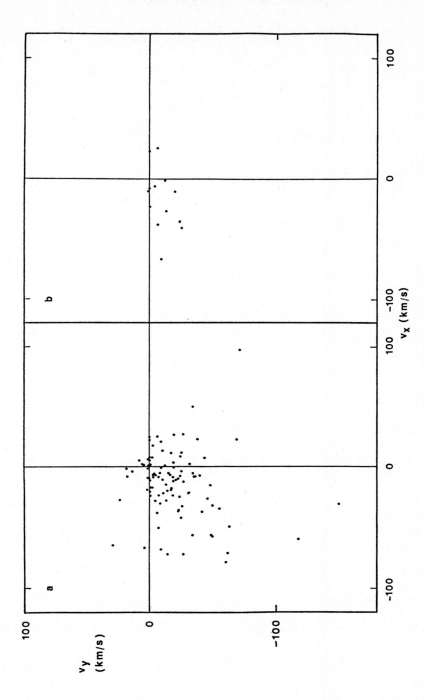

4. Plot of x and y velocity (a) for the whole sample and (b) for stars with absolute R magnitude $M_{RF} > 14$.

is the projection of $+w\sqrt{2}$ and $-u\sqrt{2}$. The reflex of the basic solar
motion is at (-11, -11). A summary of mean velocities and velocity
dispersions is given in Table 1, both with and without the 5 highest
velocity stars and it will be seen that the measures are close to the
canonical values given by Mihalas & Binney (1981). The observed solar
motion is correct to within 3 km/sec, confirming that the distances are
not seriously wrong, and the velocity dispersion is appropraite to old
M-dwarfs. Figure 4(b) shows the same diagram for stars with $M_{RF} > 14$,
and there is clearly an important change from Figure 4(a). The
velocity dispersion in the y direction is reduced to only 9 km/sec,
which suggests a very much younger population of stars. The dispersion
in the x direction is not much reduced, but a larger value is normal in
the direction of the galactic centre.

Table 1.   Velocities and velocity dispersions

	all stars	no high vel. stars	stars with $M_{RF} > 14$
n	102	97	13
$\overline{x}$	-12.2	-11.4	-16.2
$\overline{y}$	-18.6	-14.8	- 8.9
σ_x	27.3	23.6	25.1
σ_y	25.7	18.3	9.0

The first thing to be said about Figure 2(a) is that the decline in
space density from $M_{RF} = 12 - 14$ is in good agreement with that found
by Luyten, Wielen, and Reid & Gilmore for V = 12 - 15. This turnover
has in fact that modelled by Staller & de Jong (1981) and may be
associated with the cessation of hydrogen burning in stars, and so
there would appear to be no necessity to invoke a cut-off in the
stellar mass function. Of particular interest is what happends at
fainter magnitudes. Staller & de Jong's (1981) models suggest only a
rather gradual increase, while Figure 2(a) shows a steep rise in space
density at a slope roughly equal to the mass function for G-stars.
Before examining the implications of this, some discussion of the
nature of the 13 stars fainter than $M_{RF} = 14$ is appropriate. These
stars have $(R_F - I) > 2.24$, which corresponds to (R - I) > 2.01 in the
Cousins system. VB8 and VB10 are both somewhat redder than the reddest
stars in the sample and would both just lie in the faintest bin. A
spectrum of the reddest star in the sample was kindly obtained by Dr M
Dopita using the faint object red spectrograph on the Anglo-Australian
Telescope. The spectrum is illustrated in Figure 5 and is very similar
to that of LHS2930 and Wolf 359, and perhaps slightly less cool than
VB8. This fits in well with the (R - I) colour which is 2.28 in the
Cousins system. To check further the composition of the sample,
infrared colours were kindly obtained by Drs P Wood and A Longmore for
the 4 reddest sample members. The observations were made with the
infrared system on the AAT and the colours are accurate to 0.03^m. The

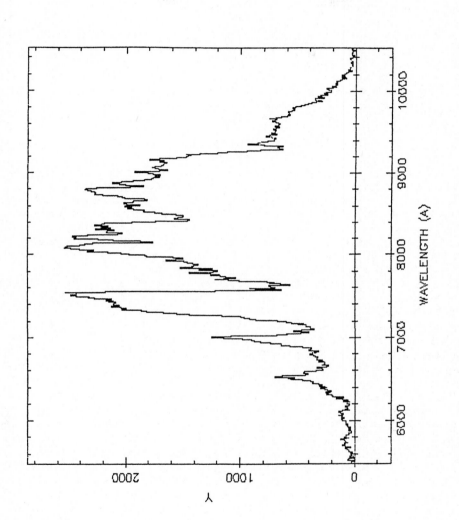

5. AAT spectrum of M01, the reddest star in the sample. The spectrum is not flux calibrated.

results, together with Cousins (R - I), are given in Table 2 and
confirm the low temperature and hence luminosity ascribed to them on
the basis of (R - I) colour.

Table 2.   Infrared colours of sample members

	R-I	K	J-H	H-K
M01	2.28	13.50	0.69	0.43
M02	2.25	13.74	0.63	0.29
M03	2.20	13.53	0.57	0.29
M04	2.17	14.12	0.62	0.28

The increase in space density fainter than M_{RF} = 14 seems to have only
one likely explanation, that we are seeing the appearance of a
population of substellar brown dwarfs in a degenerate cooling phase.
The small velocity dispersion of these faint stars suggests that they
are relatively young ($< 10^9$ years), which is consistent with
Stevenson's (1978) figures for brown dwarf cooling rates which imply
that such stars spend only a short time ($< 10^9$ years) at the
temperatures implied by the (R - I) colours of the sample. The
difference in slope between observed and predicted luminosity functions
requires further investigation, and may imply that Staller and de
Jong's (1981) cooling rates are too high. One interesting implication
of these results is that a number of well known low-luminosity stars
such as VB8 and VB10 are indeed brown dwarfs as they lie much fainter
than the break at M_{RF} = 14 which can be taken to define empirically the
cessation of hydrogen burning.

The increase in space density in Figure 2(a) clearly has important
implications for the problem of missing mass in the solar
neighbourhood. Confronted with the down turn fainter than M_V = 12, it
was perhaps natural to conclude that the space density of stars became
vanishingly small at fainter luminosities. The subsequent increase in
space density now indicates that there is indeed a steadily increasing
mass contribution at fainter luminosities. While it is true that the
13 faintest stars actually observed in the present sample do not in
themselves provide more than about 5% of the mass density required to
make up the local deficit, the upward trend in Figure 2 gives good
reason to believe that the missing mass resides in yet fainter stars.
As Stevenson (1978) has pointed out, the mass function does not have to
be extrapolated far before sufficient mass density is obtained. This
picture can only finally be confirmed by probing yet deeper into the
luminosity function. This means observing yet fainter and redder
stars, which will provide a considerable challenge as we are already
close to the limits of the UK Schmidt Telescope.

I wish to thank Dr M S Bessell for much help throughout this
investigation, and L Hartley for constructive comments on an earlier

version of the paper. I am very grateful to Drs M Dopita, A Longmore and P Wood for obtaining follow up observations for some of the stars on the AAT.

Postcript. Since this contribution was drafted an extensive recalibration of the field has been carried out. This has left the overall picture unchanged. However, the minimum now appears somewhat fainter and broader than in Figure 2, and with a less steep rise to fainter magnitudes.

References

Bessell, M.S., 1985. Preprint.
Blair, M. & Gilmore, G., 1982. Publ. astr. Soc. Pacif., 94, 742.
Hawkins, M.R.S., 1984. Mon. Not. R. astr. Soc., 206, 433.
Luyten, W.J., 1968. Mon. Not. R. astr. Soc., 139, 221.
Mihalas, D. & Binney, J., 1981. "Galactic Astronomy", Freeman, San
          Francisco.
Oort, J.H., 1965. In "Galactic Structure", p.455. eds. Blaauw, A. &
          Schmidt, M. University of Chicago Press.
Reid, N., 1982. Mon. Not. R. astr. Soc., 201, 51.
Reid, N. & Gilmore, G., 1981. Mon. Not. R. astr. Soc., 196, 15P.
Reid, N. & Gilmore, G., 1982. Mon. Not. R. astr. Soc., 201, 73.
Sanduleak, N., 1976. Astron. J., 81, 350.
Staller, R.F.A. & de Jong, T., 1981. Astron. Astrophys. 98, 140.
Stevenson, D.J., 1978. proc. Astron. Soc. Australia 3, 227.
Wielen, R., 1974. In "Highlights of Astronomy" vol. 3, p.395, Reidel,
          Dordrecht.

THE ASTROMETRIC SEARCH FOR IR DWARFS

George Gatewood, Joost Kiewiet de Jonge,
John Stein, Inwoo Han, and Lee Breakiron
Allegheny Observatory, University of
Pittsburgh, 15214

ABSTRACT
The authors are unaware of any current astrometric search
for Brown (IR) Dwarfs; the few results available have come
from studies directed toward other goals. Nevertheless, the
lack of confirmable results is somewhat unexpected. New
electronic detectors now in use at two observatories, as
well as that of planned systems, have sufficient accuracy to
detect IR dwarf companions to any of hundreds, in the latter
case millions, of stars. However, because of the limited
size of the astrometric program that can be supported by
each instrument, the number of existing and planned
astrometric facilities are insufficient for extensive
surveys.

1. DISCUSSION

The detection of specific examples of the objects known as
"brown dwarfs" (IR dwarf is a more descriptive name in
keeping with existing nomenclature) is well within the
capabilities of current astrometric technology. The first
confirmed detection of a probable IR dwarf is already
history (Harrington 1983, McCarthy 1985). Given the
accuracy of current and past astrometric programs, this
detection is not a surprise. Indeed the fact that we only
have this one confirmed detection suggests that either IR
dwarfs are less frequent than some have estimated (Bahcall
at this conference) or that they tend not to form in
binaries including a visible star.

A simple technique for ranking potential targets for
the astrometric search for planetary systems containing
Jovian type planets was presented by Gatewood (1976) and
later extended for the search for Earth-like planets
(Gatewood et al. 1980). Assuming roughly circular orbits,
the angular semiamplitude (A) of the perturbation induced in
the motion of a star by an unseen companion is dependent
upon: the orbital period (P), the mass of the companion (m),

the parallax of the system (p) and the mass of the star (M):

$$A = m\, p\, P^{2/3}\, /\, (M + m)^{2/3}\,.\qquad(1$$

Where m is relatively small, this reduces to the form given by Gatewood for planetary systems. To find a rough ranking for the various potential targets for a search for unseen objects we may set the mass of the companion and the orbital period to arbitrary standard values. This leaves free only the mass of the target and its distance. The detection index (D) is, where m and P are set to the standard values, proportional to A:

$$D =\; p\, /\, (M + m)^{2/3}\,.\qquad(2$$

Again, for planet-size companions we may neglect m. For IR dwarfs, the detection index depends slightly upon the arbitrary value chosen for the companion mass. However, if we choose a medium value for the arbitrary m, the variation is small and the derived value is sufficiently accurate for our purposes (approximately +/- 6 percent at M = 0.25 and +/- 12 percent at M = 0.08). Note that the index is larger for easier targets and scales directly with the size of the expected perturbation and that, for a given value of m, it can be determined directly from observable quantities (for main sequence stars the mass is estimated from the mass-luminosity relationship).

   The above discussion assumes that the reflected or emitted light will be much less than that from the central star.   There may be instances where this is not so, especially where the primary is a late M dwarf and the observation is made in the near IR. A good discussion of the blending effect is given by Van de Kamp (1967). We have also assumed that A was equally visible from all orientations. This is not the case if the orbit is very eccentric, but again the effect is small in all but a very few alignments of the observer and the orbital ellipse.

   The magnitude of the former effect may be estimated by finding the maximum projected diameter of ellipses of various eccentricities as viewed from random directions. In our examination of the effect, we assumed that all eccentricities were equally likely, probably a worst-case assumption. For each of 19 equally-spaced eccentricities between 0.025 to 0.975, we determined the maximum projected diameter as seen from each of ninety-four equally weighted directions.

The maximum projection (R) can be found from:

$$R = \pm A(1-E^2)[1 - \cos^2(\alpha - \theta)\cos^2(\phi)]/[1 - E\cos(\alpha)]^2 \ ^{1/2} \quad (3$$

where A is the semimajor axis of the orbit, E is its eccentricity, alpha is the angle from the semimajor axis to that yielding R, theta is the longitude of the direction from which the ellipse is viewed and phi is the latitude. Alpha can be found from the equation:

$$A \cos^2(2\ alpha) + B \cos^2(2\ alpha) + C = 0$$

where:

$$A = F^2 + G^2 \ , \quad B = -2\ G\ H \ , \text{ and } \quad C = H^2 - F^2$$

and:

$$F = (2 - E^2) \cos^2(phi) \cos^2(theta) - E^2[2 - \cos^2(phi)]$$

$$G = (2 - E^2) \cos^2(phi) \sin(2\ theta)$$

$$H = E^2 \cos^2(phi) \sin(2\ theta) \ .$$

Using equation 3) as described, we find that the effect of random eccentricity in orbits seen from random orientations introduces the predicted factor into equation 1) so that for IR dwarfs:

$$A = 0.924\ m\ p\ P^{2/3} / (M + m)^{2/3} \ . \quad (4$$

As noted, the factor is an average and there are rare situations in which eccentricity and orientation can combine to greatly diminish the likelihood of astrometric detection. However, the eccentricity must be unusually large and the unfortunate observer must view the object from a very restricted range of orientations. For example, even at the worst orientation possible, an ellipse with an eccentricity of 0.866 will still have a projected diameter 0.5 times that of the semimajor axis. As another example we note that the average projected diameter of orbits with an eccentricity of 0.975 is 0.783 times that of the semimajor axis.

"Detection" will depend upon several factors and in view of the importance of such a discovery it is probable that announcements will be made in advance of final confirmation. However, assuming that the object has been observed for a little more than one period, we may estimate the magnitude of the observed perturbation. From 4) it is evident that the largest perturbation will result from the longest period so we will assume that the period is similar to that of the period of observation. For our typical companion mass we will adopt 0.04 solar masses, a value near the mean of the range usually suggested by theorists. For the period we will adopt a decade. Then for an orbit of average eccentricity, at the average orientation to the observer, about a one-solar mass star at a distance of ten parsecs we find from 4):

A=0.924 X 0.04 X 0.10 X 4.64 /(1.0 + 0.04) =0.016 arc sec.

From 2) we see that the primary star in our example has a detection index of 0.1. Few stars within 10 parsecs will have a detection index much smaller than 0.1 while many at even greater distances will have a larger value. For example an M dwarf with a mass of 0.25 solar masses and a distance of 20 parsecs will have a detection index of 0.114 and will exhibit a perturbation of 0.020 arc seconds in the above calculation.

Thus there are several hundred stars (it so happens that these are the astrometrically best studied stars) which would, if orbited by IR dwarfs in periods of a decade or more, exhibit perturbations on the order of +/- 10 mas (milliarc second) or larger (in many cases the indicated perturbation is of the order of tenths of an arc second). Yet after decades of study by photographic astrometry, with accuracies generally sufficient to detect such variations from linear motion, we have only one confirmed detection of a probable IR dwarf. This suggests that if IR dwarfs are as common as some have suggested (for example Bahcall at this conference) they must in some manner avoid formation in wide binaries with more massive stars.

Perhaps the lack of confirmable results is responsible for the lack of active surveys for these elusive objects. No astrometric program, planned or ongoing, known to the authors is specifically designed to detect and study IR dwarfs. To date, the only astrometric results have come from standard parallax programs. This situation may change as various observatories intensify their efforts to detect extrasolar planetary systems (Gatewood et al. 1980), but astrometric discovery of IR dwarfs will in general be the result of serendipity.

Current astrometric capabilities are certainly sufficient to cover a wide range of possible IR dwarf companions. The most extensive and the most accurate photographic parallax program is that of the U.S. Naval Observatory. This is the effort responsible for the discovery noted above. The most intensive efforts of that program have obtained annual accuracies of approximately 2 mas (Harrington et al 1980) while an annual accuracy of 4 or

5 mas is more representative of the average star on that program. This is sufficient to detect a wide range of IR dwarfs orbiting any of a majority of the stars on that extensive program.

A second parallax program at the U.S. Naval Observatory employs an advanced CCD detector to reach stars of approximately the 20th magnitude. This program, though still much smaller than the other, regularly obtains accuracies of 10 mas per object per night (Monet and Dahn 1983). Long term annual accuracies will probably surpass those of the photographic program.

Another type of electronic astrometric detector, utilizing photomultiplier tubes and known as the Multichannel Astrometric Photometer (MAP) (Gatewood and Stein et al. 1980), is being employed by the Allegheny Observatory in a new and growing parallax program (Gatewood and Stein et al. 1985). The nightly accuracy of this instrument is 2 to 4 mas and, with an annual accuracy approaching 0.5 mas, should be sufficient to detect the IR dwarfs with orbits of a decade or so orbiting stars with detection indices as small as 0.005 (for example a solar mass star at 200 parsecs).

The latter two programs actually have sufficient accuracy to detect unseen objects in fairly short-period orbits. Using 4), as illustrated above, we find that assuming a 1/2-year orbit instead of a 10-year orbit about the star of detection index 0.1 results in A = +/- 2.2 mas. This suggests a considerable overlap with possible spectroscopic searches.

There are three planned astrometric instruments with still greater capacity for the detection of IR dwarfs. In order of increasing accuracy, the first is the Hubble Space Telescope planned for launch next year. This instrument is being covered elsewhere in this publication. The second, the University of Pittsburgh Astrometric Reflector to be located at the University of Hawaii's Mauna Kea site, is predicted to have an accuracy on the order of 0.1 mas. The third, the Astrometric Telescope Facility to be located on the U.S. manned space station (a University of Arizona - NASA Ames joint project in which the University of Pittsburgh is a participant) has a projected accuracy of 0.01 mas per observation.

Current and planned instrumentation have more than adequate capabilities to detect IR dwarfs if they are numerous and occur as companions to stars. If such IR dwarfs are rare, detection will probably depend more on the number of astrometric instruments in use than on the development of accuracies beyond those already envisioned. The ATF will be able to detect IR dwarfs orbiting any of millions of possible targets. However a conceivable program would examine only a few hundred regions every few years.

The accuracies of new and planned astrometric facilities are orders of magnitudes greater than those still in wide use. Unfortunately, only the accuracy of the new instrumentation is so increased, not their capability to sustain large programs. This brings about a greater need for astrometric telescopes than at any time in the past, for now we have the techniques to study whole new vistas on which the elusive IR dwarf is but one new class of object.

ACKNOWLEDGEMENTS

The authors acknowledge the continued support of the National Science Foundation, through grant #AST 8315455 and the National Aeronautics and Space Administration through grant #NAG 2-53 as well as the private contributors to the Allegeny Observatory's Endowment and new instrumentation fund.

REFERENCES

Gatewood, G. (1976). Icarus 27, 1.

Gatewood, G., Breakiron, L., Goebel, R., Kipp, S., Russell, J., and Stein, J. (1980). Icarus 41, 205.

Gatewood, G., Stein, J., DiFatta, C., Kiewiet De Jonge, J., and Breakiron, L. (1985). (1985). Astron. J. 90, 2397.

Harrington, R.S. and Dahn, C.C. (1980). Astron. J. 85, 454.

Harrington, R.S., Kallarakal, V.V., and Dahn, C.C. (1983). Astron. J. 88, 1038.

Macarthy, D.W., Probst, R.G. and Low, F.J. (1985). Astrophys J. 290, L9.

Monet, D.G. and Dahn, C.C. (1983). Astron. J. 88, 1489.

Van de Kamp, P. (1967). Principles of Astrometry (W.H. Freeman and Co. San Francisco), p.167.

THE ROLE OF SPACE TELESCOPE IN THE DETECTION OF BROWN DWARFS

William A. Baum
Lowell Observatory, Flagstaff, Arizona 86001, U.S.A.

Abstract. On reasonable assumptions, brown-dwarf companions of nearby stars should be detectable with the Hubble Space Telescope (HST) both by direct imaging and by astrometric monitoring of stellar reflex motion. If there are enough brown dwarfs in interstellar space to account for the unseen mass in the galactic disk, they should also be detectable by direct HST imaging, but identification may be tricky.

INTRODUCTION

Evidence concerning the existence and prevalence of brown dwarfs should emerge from several types of observation made with the Space Telescope. Brown dwarfs may exist as planet-like companions of stars, or they may exist as detached inhabitants of interstellar space. Those two situations require completely different observational approaches. Most of the thinking to date has been aimed at the detection of low-mass companions of nearby stars, so I shall focus much of this discussion on that problem.

Brown-dwarf companions of nearby stars will in principle be detectable with three of the instruments aboard the Hubble Space Telescope (HST):

- Wide-Field/Planetary Camera (WF/PC),
  a dual CCD camera with a choice of f/12.9 (WFC) or f/30 (PC).

- Faint Object Camera (FOC),
  a pulse-count imager using an intensifier + lens + TV tube.

- Fine-Guidance System (FGS),
  three mechanically movable star position sensors.

Details of these instruments are dealt with in "Instrument Handbooks" available from the Space Telescope Science Institute. Optical performance characteristics of the WF/PC and the FOC are summarized in Table 1.

On the basis of signal-to-noise calculations, we expect WF/PC and FOC frames to provide direct imaging detection of brown-dwarf companions in some cases (but not of "Jupiters"). However, the reflex motions of nearby stars due to companions as low in mass as Jupiter should be astrometrically detectable by the analysis of WF/PC frames obtained periodically over a number of years. The astrometric precision of the FGS is about the same as that of the WF/PC. Taken together with groundbased monitoring (radial velocities and astrometry), WF/PC and FGS astrometry should make a significant contribution. Some of us have recommended the long-term allocation of HST time for direct imaging (WF/PC and FOC) of 50 nearby stars and for systematic astrometry (WF/PC and FGS) of 20 stars. This is now a candidate Key Project as defined in the October 1985 "Call for Proposals" issued by the Space Telescope Science Institute. Initial epoch observations of a few of those stars are planned by the instrument teams.

BROWN DWARFS IN INTERSTELLAR SPACE

Brown-dwarf inhabitants of interstellar space will in principle be detectable in deep WF/PC exposures of uncluttered fields. If they exist in sufficient numbers to account for the unseen mass ("missing mass") of the Galaxy, there is a fairly good chance of discovery, unless the mass function has an unfortuitous dip in the region between 0.01 $M_\odot$ and 0.1 $M_\odot$. Based on the model of Bahcall and Soneira (1980), Bahcall estimates that 100 brown dwarfs per square degree should be within the limits of WF/PC detection,-- an average of one such object per five deep WFC (f/12.9) exposures. It seems likely that a statistically significant test for brown dwarfs in the general field can therefore be achieved with pairs of WFC frames in two colors (mainly V and I) obtained for other purposes by GTO observers during the first year of Space Telescope observation. However, the image analysis will not be easy. Not only will most WFC frames be cluttered with

Table 1. Hubble Space Telescope Cameras.

	WF/PC		FOC			
F-Ratio	F/12.9	F/30	F/48	F/96	F/288	-
Format	2×800	2×800	512	512	512	Pixels
Field	154	66	22	11	3.6	Arcsecs
Scale	0.10	0.043	0.043	0.022	0.0072	Arcsec/px
λ Range	0.12-1.1	0.12-1.1	0.12-0.6	0.12-0.6	0.12-0.6	μm
* Range	11-28	10-28	22-27	21-27	18-27	V mag
DQE　0.3μ	0.06	0.06	0.09	0.13	0.09	-
DQE　0.7μ	0.16	0.16	0	0	0	-

Formats and fields are all squares having the side-dimensions indicated. Fields and image scales are for standard photometric imaging modes. Wavelength range is for detective quantum efficiencies (DQE) > 0.01. Stellar magnitude range is for wide-band filters, S/N ~ 5, t ≤ 2500 sec. DQE includes telescope throughput but not filter transmissions.

stellar images that are not brown dwarfs, but there will also typically be more than 1000 extragalactic objects (many very red and nearly stellar in appearance) in each WFC frame. Finding the brown dwarfs will therefore be a needle-in-the-haystack problem, and any discovered candidates will almost certainly be equivocal cases with varying degrees of likelihood. Individual candidates will then have to be investigated further with additional color data and with spectra.

DIRECT IMAGING DETECTION OF STELLAR COMPANIONS

If brown dwarfs are found by direct HST imaging to be companions of nearby stars, we will learn their individual absolute magnitudes and (in some cases) their colors. Such data will be extremely helpful in constraining theoretical models. Given the substantial percentage of stars already known to have stellar companions, and given the assortment of planets associated with the Sun, it would be quite remarkable if bodies in the mass range of brown dwarfs did not also occur as stellar companions.

To evaluate the HST detectability of brown-dwarf companions of nearby stars by direct imaging, let's look at two strawman cases, one with a mass large enough (0.05 $M_\odot$) for gravitational contraction to produce substantially more radiation than the light reflected from the primary star, and the other of small enough mass for the contribution of gravitational contraction to be unimportant. It is not clear to me from the published literature what the photometric parameters of a brown dwarf of 0.05 $M_\odot$ are, so please forgive me if I now invent one that doesn't accord with your favorite view.

Based on van der Linden and Staller (1983), a 0.05 $M_\odot$ brown dwarf should be around 20th absolute visual magnitude at some stage in its life. Suppose it to be in a circular orbit about like that of Jupiter around a solar-type star. Specifically, to play a game with some numbers, put the brown dwarf 5 AU from the star (a rather tough case) and put the star 5 parsecs from us. If viewed near maximum elongation, the dwarf would then be an 18.5-magnitude (visual) companion, 1 arcsecond from a 3.3-magnitude G-type star. In the infrared, however, the magnitude difference would be smaller. For the wide-I passband of the WF/PC, our hypothetical brown dwarf would be a 14th magnitude object to be detected against a 12th magnitude patch in the wing of the point-spread function of a 2.6-magnitude star.

Let me clarify the concept of a background patch. For signal-to-noise calculations, there exists an effective area of background (i.e., an effective number of pixels) against which the point-spread function of a faint object inherently competes. For the wide-I passband and for the f/30 PC, this background patch turns out to contain 26 pixels, and a patch of that size located 1 arcsecond from a 2.6-magnitude star (I band) amounts to 12th magnitude.

To avoid severe saturation of the CCD, the image of the 3rd magnitude parent star can be positioned on an attenuating spot in the field of the

WF/PC, known whimsically as the "Baum Spot". This spot, which is 1.4 arcseconds in diameter, was created with the brown-dwarf/ planet search in mind. The spot will attenuate the parent star image about 7 magnitudes. Even so, the exposure time cannot be more than about 1 second without producing an objectionable amount of charge overflow into pixels neighboring the core of the star image. But in a single 1-second exposure, our postulated brown dwarf would be detected with a signal-to-noise ratio greater than 10. A piece of cake! And for larger separations between the star and the brown dwarf, detection would be even easier.

But now let's look into direct imaging detection of a less massive brown dwarf, say 0.005 $M_\odot$. That is much more difficult. It would require the stacking of a very large number of separate 1-second exposures, a technique that improves the signal-to-noise ratio in proportion to the square-root of the number of images stacked. Treating this 0.005 $M_\odot$ dwarf as a super-Jupiter belonging to a solar type star at 5 parsecs, we find it to have V = 25.2 and I = 24.5 magnitude at greatest elongation. If observed with the PC in the V passband, for which the effective number of pixels in the background patch is 12, the background (for calculating the signal-to-noise ratio) is V = 14.4 magnitude. It turns out that for bare detection (S/N = 5), we would need about 200 hours of accumulated exposure time, formed (for example) by stacking about 720,000 exposures of 1 second each. I hardly need to explain why that is impractical. The situation is not dramatically better for other assumed separations. The bottom line: <u>evolved</u> companions that are not self-radiating will not be detected by direct WF/PC imaging.

The environs of very <u>young</u> stars may, however, include large <u>unevolved</u> proto-planets as well as the the circumstellar disks recently discovered; direct imaging of them with the Space Telescope could yield interesting results.

One strategy for direct HST imaging of the environs of a nearby star will be to use the Faint Object Camera (FOC) in its f/288 coronagraphic mode, so that the star is occulted by a 0.8-arcsecond finger and the diffraction is suppressed by a pupil-plane apodizer. Even so, very long integrations would be required to build up the signal-to-noise ratio for features that are to be detected against the residual scattered starlight. It is not yet clear whether the FOC coronagraph will excel the PC + Baum Spot for this type of observation.

ASTROMETRIC DETECTION OF STELLAR COMPANIONS

To provide a feel for the amount of stellar reflex motion that low-mass companions may be expected to cause, Figure 1 shows the actual wobble of the Sun due to our own planetary system, plotted for a 100-year interval (1980 to 2080). The contribution of Jupiter alone is 0.01 AU peak-to-peak. The amplitude will scale almost linearly with companion mass and separation, so a brown dwarf of 0.05 $M_\odot$ at Jupiter's orbital distance (for example) would produce 0.5 AU peak-to-peak.

Two instruments aboard HST having the desired astrometric capability
(Baum 1980) are the FGS and the WF/PC. Both are capable, under some
circumstances, of detecting positional displacements of a program star
with respect to its neighbors with a precision of roughly 0.002 arcsec
(RMS) for single observations. The two instruments differ substantially,
however, in their stellar magnitude limits and in their requirements
concerning the distribution of reference stars.

The FGS has three movable probes that can be positioned to acquire stars
within three pickle-shaped areas around the periphery of the ST focal
plane. While two probes are locked onto distant reference stars, the
third can take an astrometric reading on the position of a nearby
program star. For high astrometric accuracy, all of those stars should
be brighter than about 14th magnitude. Performance is degraded if one of
them turns out to be a close double.

The WF/PC (as mentioned before) is a dual-mode camera. In each mode,
four 800×800-pixel CCDs provide a 2×2 mosaic of the image field, but for
best long-term results, we are interested in only the quarter of the
image field that falls within a single CCD. If the program star needs to
be attenuated with respect to surrounding reference stars, one would
choose the particular CCD on which the Baum Spot falls. The field of a
single CCD in the f/12.9 WFC is about 77×77 arcseconds, while that of a
single CCD in the f/30 PC is 33×33 arcseconds. Astrometric centroiding
precision is slightly higher with the PC than with the WFC, but the WFC

Figure 1. Reflex motion of the Sun over a 100-year interval
due to the orbiting of major planets, as it would be seen by
a distant observer at RA = 0^h, DEC = 0°. The amplitude
(ordinate) is in AU.

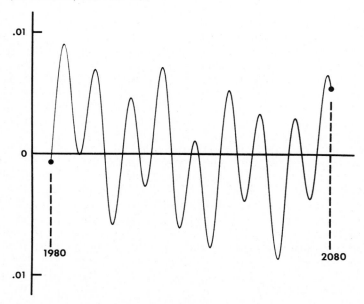

provides a fivefold larger field area of reference stars, so there is a tradeoff that depends on the character of individual star fields. For precision astrometry, the dynamic range is 3 to 4 magnitudes (plus an offset of 7 magnitudes for any star falling within the Baum Spot), but the magnitude limit depends on the exposure time and filter selected. The faintest stars that can be accurately dealt with astrometrically by the WFC in a single orbital night (2500 sec) are about 24th magnitude! This means that nearby stars as bright as 14th magnitude can be monitored astrometrically utilizing comparison stars as faint as 24th magnitude. Similarly, a program star of 10th magnitude can be astrometrically measured against comparison stars down to 20th magnitude in an exposure time of one minute.

SELECTION OF STARS TO BE OBSERVED
        For direct imaging, we are interested in surveying two classes of targets: (a) stars that are unusually nearby,. and (b) young stars that may have planetary disks in various stages of evolution.

For the nearby stars, priority should be given to those within 5 parsecs. About 50 (including companions in double systems) are now known; they are listed in Table 2. Their seasons of HST accessibility are spread throughout the year, but some are easier for HST cameras to deal with than others. The selection of an actual observing list will be a complex process taking account of the magnitude, color, and environment of each star, together with performance data for the FOC and the PC. If the survey radius is extended beyond 5 parsecs, the number of candidates increases rapidly, but it does not increase as fast as the cube of the survey radius because of the increasing incompleteness of catalogs. Images of a few of these nearby stars will be obtained by the Guaranteed Time Observers (GTOs) during early stages of the HST mission.

Among these very nearby stars, there are not enough young ones suitable for investigating the evolution of proto-planets and planetary disks, so a list of disk candidates has to be drawn up on different criteria. Preparations should include a systematic review of IRAS data and a systematic program of ground-based imaging with a coronagraphic CCD camera. A few planetary disk candidates will be imaged with HST cameras during GTO time, and those early results will help determine how extensive a disk survey should be carried out with HST in subsequent work.

With regard to astrometric candidates, we have a choice of several different selection criteria. Nearness (per Table 2) is a factor, but not the whole story. To develop one criterion, suppose each star to possess a hypothetical "Jupiter," defined as a planet of 0.001 solar mass located in a quasi-circular orbit with a semi-major axis of 5 AU. Stars are listed in Table 3 in descending order of "alpha," the semi-amplitude (in milliarcseconds) of reflex motion due to this hypothetical Jupiter. There are 40 stars for which alpha exceeds 5 milliarcseconds. For a brown dwarf, we can simply express both mass and orbital radius in Jupiter units, and multiply alpha by the product. Unfortunately, many of

Table 2. The nearest stars. See text for meaning of α and period P.

GLIESE	NAME	RA	DEC	π	α	P
551	PROXIMA CEN	14 26 19	-63 16	0.763	36.118	36.6
559A	ALF CEN	14 36 12	-60 4	0.741	2.721	10.2
559B	(RIGIL KENT)	14 36 12	-60 4	0.741	3.704	11.9
699	BARNARD STAR	17 55 24	4 26	0.552	15.751	28.4
406	WOLF 359	10 54 5	7 20	0.426	26.726	42.2
411	DM+36 2147	11 0 36	36 15	0.397	5.992	20.7
244B		6 42 58	-16 40	0.377	7.290	23.4
65A	L 726-8	1 36 24	-18 13	0.368	16.689	35.9
65B	UV CET	1 36 24	-18 13	0.368	18.855	38.1
729AC	-24 2833-183	18 46 46	-23 48	0.345	9.945	28.6
905	ROSS 248	23 39 26	43 58	0.318	12.974	34.0
866	L 789-6	22 35 46	-15 34	0.305	11.861	33.2
144	EPS ERI	3 30 34	-9 34	0.302	1.672	12.5
447	ROSS 128	11 45 10	1 2	0.301	9.096	29.2
820A	61 CYG	21 4 41	38 25	0.296	2.286	14.8
820B		21 4 41	38 25	0.296	2.755	16.2
845	EPS IND	21 59 34	-56 51	0.291	1.965	13.8
280B		7 36 41	5 24	0.285	7.668	27.6
15A	DM+43 44	0 15 31	43 38	0.282	4.090	20.3
15B	DM+43 44	0 15 31	43 38	0.282	8.105	28.5
725A	DM+59 1915	18 42 12	59 32	0.282	4.952	22.3
725B		18 42 12	59 32	0.282	5.940	24.4
887	DM-36 15693	23 2 38	-36 7	0.279	3.429	18.6
71	TAU CET	1 41 46	-16 15	0.277	1.395	11.9
273	LUYTENS STAR	7 24 43	5 26	0.270	5.752	24.6
825	DM-39 14192	21 14 19	-38 57	0.260	2.627	16.9
191	KAPTEYEN STAR	5 9 41	-44 54	0.256	4.196	21.6
860A	DM+56 2783	22 26 12	57 28	0.253	5.253	24.2
860B	DO CEP	22 26 12	57 28	0.253	7.301	28.6
234A	ROSS 614	6 26 50	-2 45	0.252	6.906	27.9
628	DM-12 4523	16 27 31	-12 30	0.249	5.442	24.9
35	VAN MAANEN 2	0 46 31	5 13	0.239	8.606	31.9
473A	WOLF 424	12 30 50	9 18	0.231	9.806	34.7
473B		12 30 50	9 18	0.231	10.316	35.6
1	DM-37 15492	0 2 29	-37 37	0.225	3.324	20.4
380	DM+50 1725	10 8 19	49 51	0.222	2.036	16.1
674	DM-46 11540	17 24 53	-46 53	0.216	3.693	22.0
832	DM-49 13515	21 30 14	-49 17	0.214	3.109	20.3
83.1	L 1159-16	1 57 29	12 49	0.213	7.077	30.7
682	DM-44 11909	17 33 29	-44 23	0.213	5.481	27.0
687	DM+68 946	17 36 41	68 23	0.213	3.450	21.4
876	DM-15 6290	22 50 36	-14 32	0.209	4.242	24.0
440	L 145-141	11 42 58	-64 34	0.206	5.562	27.6
166A	KEID	4 12 58	-7 46	0.205	1.098	12.3
166B	DM- 7 781	4 13 5	-7 46	0.205	3.553	22.2
166C	DM- 7 781	4 13 5	-7 46	0.205	5.183	26.8
388	DM+20 2465	10 16 53	20 11	0.204	3.450	21.9
526	DM+15 2620	13 43 12	15 8	0.201	2.721	19.6
445	AC+79 3888	11 44 36	79 13	0.196	4.575	25.7
702B		18 2 55	2 27	0.195	1.462	14.6

Table 3. Nearby stars ordered according to reflex semi-amplitude α.

GLIESE	NAME	RA			DEC		π	α	P
551	PROXIMA CEN	14	26	19	-63	16	0.763	36.118	36.6
406	WOLF 359	10	54	5	7	20	0.426	26.726	42.2
65B	UV CET	1	36	24	-18	13	0.368	18.855	38.1
316.1	LP 425-140	8	37	41	23	52	0.139	18.335	61.1
752B	VB 10	19	14	31	5	4	0.173	16.791	52.4
65A	L 726-8	1	36	24	-18	13	0.368	16.689	35.9
699	BARNARD STAR	17	55	24	4	26	0.552	15.751	28.4
283B		7	38	2	-17	16	0.142	13.256	51.4
905	ROSS 248	23	39	26	43	58	0.318	12.974	34.0
644C	VB 8	16	52	55	-8	20	0.161	12.762	47.4
866	L 789-6	22	35	46	-15	34	0.305	11.861	33.2
473B		12	30	50	9	18	0.231	10.316	35.6
729AC	-24 2833-183	18	46	46	-23	48	0.345	9.945	28.6
473A	WOLF 424	12	30	50	9	18	0.231	9.806	34.7
412B	WX UMA	11	3	2	43	45	0.186	9.710	38.5
447	ROSS 128	11	45	10	1	2	0.301	9.096	29.2
35	VAN MAANEN 2	0	46	31	5	13	0.239	8.606	31.9
293	L 97-12	7	52	48	-67	46	0.173	8.281	36.8
223.2	LP 658-2	5	52	38	-4	11	0.166	8.173	37.3
15B	DM+43 44	0	15	31	43	38	0.282	8.105	28.5
618	B	16	16	48	-37	27	0.131	8.081	41.8
280B		7	36	41	5	24	0.285	7.668	27.6
860B	DO CEP	22	26	12	57	28	0.253	7.301	28.6
754	L 347-14	19	17	7	-45	36	0.175	7.301	34.4
300	L 674-15	8	10	29	-21	21	0.171	7.292	34.8
244B		6	42	58	-16	40	0.377	7.290	23.4
83.1	L 1159-16	1	57	29	12	49	0.213	7.077	30.7
234A	ROSS 614	6	26	50	-2	45	0.252	6.906	27.9
169.1B	AC+58 25002	4	26	48	58	53	0.192	6.297	30.5
518	WOLF 489	13	34	12	3	57	0.135	6.254	36.2
411	DM+36 2147	11	0	36	36	15	0.397	5.992	20.7
725B		18	42	12	59	32	0.282	5.940	24.4
102	L 1305-10	2	30	43	24	41	0.133	5.805	35.2
693	L 205-128	17	42	24	-57	20	0.170	5.763	31.0
273	LUYTENS STAR	7	24	43	5	26	0.270	5.752	24.6
440	L 145-141	11	42	58	-64	34	0.206	5.562	27.6
682	DM-44 11909	17	33	29	-44	23	0.213	5.481	27.0
628	DM-12 4523	16	27	31	-12	30	0.249	5.442	24.9
860A	DM+56 2783	22	26	12	57	28	0.253	5.253	24.2
166C	DM- 7 781	4	13	5	-7	46	0.205	5.183	26.8
725A	DM+59 1915	18	42	12	59	32	0.282	4.952	22.3
299	ROSS 619	8	9	12	9	2	0.151	4.733	29.8
169.1A	AC+58 25001	4	26	48	58	53	0.192	4.615	26.1
445	AC+79 3888	11	44	36	79	13	0.196	4.575	25.7
288B	VB 3	7	43	50	-33	48	0.063	4.565	45.3
283A	L 745-46	7	38	2	-17	16	0.142	4.554	30.1
432B	VB 4	11	32	5	-32	36	0.105	4.481	34.8
783B		20	7	55	-36	14	0.177	4.446	26.7
359	ROSS 92	9	38	12	22	16	0.092	4.403	36.8
213AC	12 1800-213	5	39	14	12	31	0.168	4.251	26.8

the orbital periods P turn out to be long compared with the planned
lifetime of HST; for those cases, too small a fraction of one cycle of
the reflex motion is observed for us to be sure of distinguishing it
from linear proper motion.

This brings us to a second selection criterion: we should prefer
candidates that are expected to have the <u>greatest departure from linear
proper motion</u> within a limited observing interval (say 10 to 15 years).
Specifically, they are the ones having the highest second derivative of
reflex displacement. That leads to the fascinating result that the peak
departure from linear proper motion, expressed in milliarcseconds/year^2,
is simply 1.4 times the parallax in arcseconds. Thus, alpha and parallax
are both important in the case of HST. Those two parameters are plotted
in Figure 2, where a few of the most favorable candidates are labeled.

Figure 2. Candidate stars for astrometric monitoring with HST for the
purpose of detecting reflex motion due to the orbiting of low-mass companions. Alpha is the semi-amplitude that would be produced by a body of
0.001 $M_\odot$ at 5 AU. For both alpha and parallax, the larger the better.

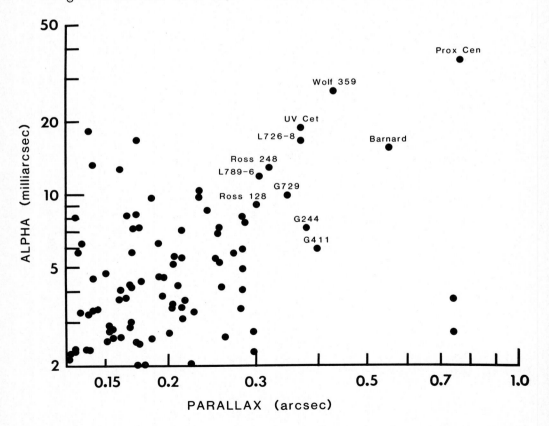

A third criterion, suggested by Gatewood (1976), is the parallax divided by the two-thirds power of the mass of the star. Although this is called a "detection index", the meaning in terms of companion detectability is not directly clear.

Some of the stars in Figure 2 (not necessarily just those labeled) will be observed during GTO time with the FGS, some with the WF/PC, and a few with both. The point of planning such observations under a long-term HST project is to ensure that the _same_ stars will continue to be observed regularly with the same instruments for the longest possible time span.

REFERENCES

Bahcall, J.N. & Soneira, R.M. (1980). The universe at faint magnitudes. I. Models for the Galaxy and the predicted star counts. Astrophys.J.Suppl., 44, 73-110.

Baum, W.A. (1980). The ability of the Space Telescope to detect extra-solar planetary systems. Celestial Mechanics, 22, 183-190.

Gatewood, G. (1976). On the astrometric detection of neighboring planetary systems. Icarus, 27, 1-12.

van der Linden, T. & Staller, R. (1983). Evolution of very low mass stars. Astron.Astrophys., 118, 285-288.

II. THEORY

The papers that follow present detailed models of the structure of brown
dwarfs (BDs). The first theoretical models about the onset of electron
degeneracy resulting in the termination of the collapse of low mass protostars were developed in the early 60's. The term "brown dwarf" itself
was devised in the 70's. Recent models indicate that there is a boundary
class in which objects are partially supported by nuclear burning and
partially by thermal cooling. VB8 B may indeed be such an object. Cooling
curves for brown dwarfs have been calculated resulting in general agreement between the suspected age of VB8 B and the cooling time of BDs
($\lesssim$ a few $x10^9$ yr). Specific models follow the evolution of low mass stars
($M \lesssim 0.1$ $M_\odot$) through deuterium burning to the very late stages of degenerate cooling. Protostars as low as ~ 0.02 $M_\odot$ can form in fragmentation
of Population I clouds and low metallicity may help the fragmentation of
substellar objects. Planetary interior physics have been extended to
higher densities and pressures applicable to the BD regime and the relation of Jupiter-size objects to BDs examined. It is found that the shape
of the infrared spectrum depends on the uncertain envelope composition.
The main uncertainty in the theoretical models of BDs to this date remains
the infrared opacity.

The general consensus seems to be that objects in the approximate range
0.01 - 0.08 $M_\odot$ with cooling timescales $\lesssim$ a few $x10^9$ are brown dwarfs.
Also the consensus seems to be that the mass of VB8 B is probably 0.06 $\pm$
0.02 $M_\odot$. The main challenge for the future may lie in the identification
of more BDs and for this it is clear that spectral information is needed.
The mass function of BDs may be very difficult to obtain from current
observations making a direct estimate of the contribution of BDs to the
local mass density difficult indeed. However, BDs could indeed provide
the main contribution to the unseen matter with local mass density of
~ 0.1 $M_\odot$ pc^{-3}. A variety of arguments were presented favoring this point
including the position of BDs on diagrams that include a large number of
other astronomical objects. BDs could be quite numerous and relatively
nearby, maybe as close as 0.5 pc from the sun. In hindsight, it appears
that at those close distances they would have profound effects on the
Oort cometary cloud, perhaps triggering events that involve comet showers
every few million years that may result in biological extinctions.

AN HISTORICAL PERSPECTIVE: BROWN IS NOT A COLOR

J.C. Tarter
Astronomy Department, University of California Berkeley
and
SETI Institute

Abstract In the twelve years since I started working with Professor Joe Silk on the brown dwarf part of my thesis, the astronomical community has come full circle in its thinking about these objects. Back then, Brown Dwarfs with masses down to planetary values could not be excluded by any theories of fragmentation, indeed they seemed the inevitable result of two unrelated bits of physics. These objects were incredibly useful in explaining various "missing mass" problems, particularly within the local solar neighborhood, but unfortunately the instrumental sensitivity was so many of orders of magnitude away from being able to detect them, that they were for all practical purposes invisible. Today, Larson's refinements to the theory of hierarchical fragmentation may well exclude the possibility of the formation of at least the lower end of the mass range for such objects. The "missing mass" in the local solar neighborhood might be satisfactorily explained by remants of stars at the other extreme end of the initial mass function, in which case any substantial contribution to the local density due to brown dwarfs would be an embarrassment. Nevertheless, instrumental capabilities have improved so rapidly, that we may actually have observed one! In this paper I shall try and trace the major shifts in our theoretical understanding of the star formation process and the possible components of the local mass density. I shall also outline those aspects of Brown Dwarf structures and evolution that are still not well enough understood and suggest the type of observations that might force us to modify our current theories to once again accomodate the existence of these objects. Along the way I hope to defend the term I coined and convince you that such hypothetical objects which are definitely not stellar, and that may not be related to the gas giant planets, and whose gross emission spectra cannot be predicted should properly be called by a name that is not a color.

INTRODUCTION

For me the "present" is defined by the papers currently covering the top of my desk and the "past" is defined by what is in my filing cabinets. Therefore, having no particular talents as a historian, I shall present below a biased historical perspective based entirely on the content of one of my filing cabinets! Each section heading corresponds to an actual label on a file folder as preserved in that cabinet. These headings should be pardoned for their colloquial nature as they were never intended for publication. I apologize in advance for any omissions in this summary and would be pleased to be informed about any such oversights.

"LOW MASS STARS" \ "STAR FORMATION" \ AND / "MISSING MASS"

These three folders represent a large collection of reference papers gathered together in the early 1970's in order to try to answer a question posed by my thesis advisor. Joe Silk asked "Are there any observational constraints that can be placed on the hypothesis that the "missing mass" in the Coma Cluster can be explained by many stars too small to stably burn hydrogen and therefore too dim to be observed?"

Kumar (1963 a,b) recognized that the onset of electron degeneracy and the resultant increase in pressure would terminate the collapse of a low mass protostar at a central temperature that was too low to support a stable hydrogen fusion process. He proposed that the zero age main sequence must therefore truncate near $0.1 M_\odot$. Thermal models for Jupiter (Peebles 1964, Hubbard 1968, Hubbard & Slattery 1971 and Salpeter 1973) were available to model the slow gravitational cooling of objects that failed to stabilize on the main sequence. In addition cooling-at-constant-radius (Ostriker & Axel 1969) models successfully predicted the luminosity of white dwarfs cooling to invisible black dwarfs characterized by the zero temperature equilibirum configuration of Salpeter & Zapolsky (1967). The equation of state for high density, low temperature, primordial abundance interiors of such objects was sufficiently well understood to promote a large body of numerical calculations to determine the exact mass cut-off to the main sequence. Most notable were a series of papers by Grossman, Graboske and co-workers (Grossman 1970, Grossman et al 1970, Grossman & Graboske 1971, and Grossman et al 1974). The lower limit to the main sequence was pegged at $0.085 M_\odot$ with (then as now) the major uncertainties being attributed to the need to extrapolate the low Z opacities (Cox & Stewart 1970) into density and temperature regimes where they were poorly known.

The question of whether any objects of mass lower than this limit actually existed was approached both theoretically and observationally. Based on the original star formation work of Hyashi (1962) and Field (1970) theorists considered the minimum mass that could form in a hierarchial, opacity limited, fragmentation process (with and without the grain opacities of Gaustad (1963)) and derived limits ranging from $0.002 M_0$ (Larson 1973) to $0.03 M_\odot$ (Lynden-Bell 1973). Kumar (1972) avoided any theoretical uncertainties by relying on early mass estimates for dark companions in multiple systems. Subsequent improvement in the estimates left only 3 candidates, Ross 614B (Lippincott & Hershey 1972) Wolfe 424A and Wolfe 424B (Heintz 1972) for objects with masses below $0.085 M_\odot$. In all cases the errors were sufficiently large to allow for the possibility that these were indeed actually faint stars. Observational data on the colors, temperatures and luminosities of the faintest red dwarf stars (Luyten 1968, Weistrop 1972 and Veeder 1974) was not sufficiently precise to distinguish between the stellar and substellar possibilities.

The contents of these folders did not prove the existence of objects less massive than hydrogen burning stars, but there was nothing to contradict the idea. As to how many such objects there might be, the most natural assumption was that, having formed by the same fragmentation process as normal stars, they would share the same initial mass function that had been attributed to the stars. The most widely used functional form being the power law distribution of Salpeter (1955).

"THESIS"

Not surprisingly, this folder contains a copy of my 1975 Ph.D. thesis from U.C. Berkeley, the fourth chapter of which is entitled "Brown Dwarf Stars and How They Grew Old". This marks the first written use of the term brown dwarf and it derived from a conversation between Joe Silk and myself in the library on the 5th floor of Campbell Hall soon after I began to try and answer the question Joe had originally posed. The terms white dwarf, red dwarf and black dwarf all existed in the astronomical vocabulary. But they referred specifically to the end products of low mass stellar evolution, to low mass main sequence stars and to the end products of white dwarf cooling with a mean molecular weight that is characteristic of billions of years of nucleosynthesis. It was obvious that we needed a color to describe these dwarfs that was between red and black. I proposed brown and Joe objected that brown was not a color. As I already understood that the inadequacies of existing opacity tables and numerical codes would prevent us from deriving a meaningful color temperature for these objects, brown, a non-color, seemed to me to fit our needs precisely. I am startled by the similarity between our conversation and the one amongst the astrometrists several years earlier reported by Bob Harrington in his presentation at the start of this workshop. Had that conversation taken place somewhere other than a bar, the world might have known about brown dwarfs a bit sooner!

"1976 BD MANUSCRIPT"

In 1976 I submitted an overly long manuscript, too much in the style of a thesis, to the Astrophysical Journal for publication. This paper was rejected by a referee as being overly long and too much in the style of a thesis and not very interesting anyway. Except for the numerous times this manuscript has been removed for reproduction, it has remained buried within its folder in my file cabinet. For comparison with later results, I take this opportunity to summarize the main points in that manuscript.

The primary question was what, if any, observational constraints could be placed on the assertion that a population of brown dwarfs could account for the "missing mass" within clusters of galaxies, massive haloes surrounding spiral galaxies and in the local solar neighborhood. Since brown dwarfs would be most observable during the early phase of their cooling, the analytic cooling-at-constant-radius models were not appropriate to describe the temporal evolution of the luminosity and color of the ensemble population. Instead empirical methods were used to extrapolate into the future unpublished numerical results of Grossman and Graboske. These had been computed for .015 $M_\odot$ and .02$M_\odot$ models and evolved beyond the time when non-equilibrium nuclear fusion ceased to contribute to the luminosity. In analogy with white dwarf cooling models, a "cooling time" was defined as the length of time required for the central temperature to drop below the Debye temperature. Because only 2 of the numerical models had been sufficiently evolved, it was not possible to solve for this cooling time self consistently. Instead the time for the central temperature to drop below the Debye temperature appropriate to the central density of the zero temperature Salpeter and Zapolsky equilibrium state i.e. $T_D(\rho_0)$ was taken as a measure of the actual cooling time with a caution that this would be an underestimate.

The resultant time dependences of the physical parameters of the brown dwarf and the cooling time were:

$$\frac{L}{L\odot} = 4.7\text{x}10^{-2}\left(\frac{m}{M\odot}\right)^{2.2}\left(\frac{t}{10^9\text{yr}}\right)^{-0.84} \text{ for } T_C(t) \geq T_D(\rho 0) \tag{1}$$

$$= 0 \text{ for } T_C(t) < T_D (\rho 0) \tag{2}$$

$$\frac{R}{R\odot} = 3.8\text{x}10^{-2}\left(\frac{m}{M\odot}\right)^{-0.22}\left(\frac{t}{10^9\text{yr}}\right)^{-0.1} \tag{3}$$

$$\frac{t_{cool}}{10^9\text{yr}} = 6.1\text{x}10^8\left(\frac{m}{M\odot}\right)^{1.2}\left(T_D(\rho 0)\right)^{-1.3} \tag{4}$$

Assuming that brown dwarfs radiate like black bodies (the only reasonable assumption in the absence of detailed opacities) led to an effective temperature that decreased like

$$T_e = 1.42\text{x}10^4\left(\frac{m}{M\odot}\right)^{0.66}\left(\frac{t}{10^9\text{yr}}\right)^{-0.16} \text{ in } {}^{\circ}K \tag{5}$$

Modifying the formalism introduced by Tinsley (1973) to represent the total mass and time dependent luminosity of an ensemble of stars and brown dwarfs (both described by the same initial mass function [IMF] and birthrate function) allowed for comparison with various observational constraints. The IMF was assumed to be a power law $\frac{dn}{dm} \propto m^{-x}$;

1.35 < x < 3) and the birthrate function was allowed to be an initial burst, constant or exponentially decaying in time. For each postulated "missing mass" environment, the IMF was continued to a brown dwarf mass low enough to provide the requisite total mass. The time dependent luminosity and colors of the visible stars and cooling brown dwarfs were compared to available data. Given the formal suppression of brown dwarf luminosity after the relatively short calculated cooling times ($2.9\text{x}10^9$ years for .08$M_\odot$ to $1.4\text{x}10^9$ years for .003$M_\odot$) it should be no surprise

that only the largest values of x combined with a constant birthrate
function did violence to any observational limits.

Figure 1 summarizes the observability of brown dwarfs from these early
calculations. Since the brown dwarfs spend the majority of their
cooling phase near the limiting luminosity at t_{cool}, the most likely
place to find an individual brown dwarf is near the curve marked
L (m, t_{cool}). Considering the local solar neighborhood and using a disk
population of brown dwarfs to provide the missing mass led to an average
luminosity and an average effective temperature lying within the shaded
region marked $\langle L, T_e \rangle$ for various combinations of IMF and birthrate
function. The solid curves labeled with model masses represent the
numerical calculations of Grossman, Graboske and co-workers and the
dashed extensions are the empirical cooling curves. The triangles refer
to model ages of 10^8 years and the stars represent the positions of the
low mass stars listed in the figure, given the estimates for red dwarf
bolometric corrections being used circa 1976. From these curves VB8B
would most likely be a $.02 M_\odot$ brown dwarf of age 2×10^8 years.

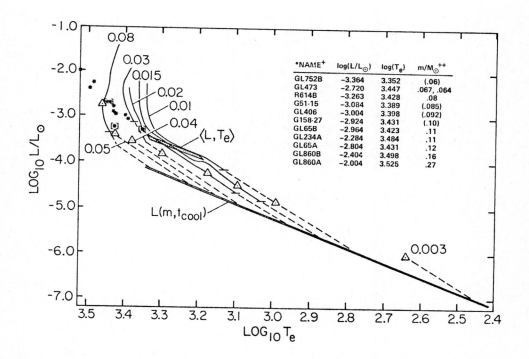

FIGURE 1

"BROWN DWARFS REVISITED"

The referee's comments not withstanding, the existence of brown dwarfs and their relationship to "missing mass" was indeed an interesting problem that was explored by a number of individuals over the next few years. Stevenson (1978) approached the problem analytically using a cooling-at-constant-(zero temperature equilibrium)-radius approximation for the degenerate cooling curve to conclude that

$$\frac{L}{L\odot} = 9\times10^{-5} \left(\frac{m}{M\odot}\right)^{1.93} \left(\frac{R}{R\odot}\right)^{-1.87} \left(\frac{t}{10^9 yr}\right)^{-4/3} \tag{6}$$

with a constant $\dfrac{R}{R\odot} = 0.105 \left[\dfrac{2\left(\dfrac{m}{.0032 M\odot}\right)^{1/4}}{1 + \left(\dfrac{m}{.0032 M\odot}\right)^{1/2}}\right]^{4/3}$ $\qquad$ (7)

Stevenson and Salpeter (1977) (also Stevenson 1979) realised that objects with masses similar to the giant planets could have an additional source of internal luminosity. This luminosity would result from the immiscibility of He in metallic Hydrogen and its subsequent migration to the core of the body, creating a more centrally condensed configuration and releasing additional gravitational energy.

Black (1980) combined numerical models of gas giant planets with the low mass star models of Grossman, Graboske and co-workers to produce semi-empirical cooling curves for planetary mass objects that predicted

$$\frac{L}{L\odot} = 0.12 \left(\frac{m}{M\odot}\right)^{2.35} \left(\frac{t}{10^9 yr}\right)^{-1.22} \tag{8}$$

$$\frac{R}{R\odot} = 0.16 \left(\frac{m}{M\odot}\right)^{0.16} \left(\frac{t}{10^9 yr}\right)^{-0.05} \tag{9}$$

$$T_e = 6\times10^3 \left(\frac{m}{M\odot}\right)^{0.51} \left(\frac{t}{10^9 yr}\right)^{-0.28} \ {}^{\circ}K \tag{10}$$

This behavior is intermediate between Tarter (1975) and Stevenson (1978), as appropriate to older configurations, that are still not yet on the constant radius asymptote. It is unfortunate that Black's cooling curve received little attention for it might have provided theorists with the best representation of the evolution of a brown dwarf that is neither too young and luminous, nor too old and faint to be representative. Both Tarter and Stevenson had concluded that if brown dwarfs accounted for the local "missing mass", then the average space density of these objects should be ~ few per pc^3: the main question was what their expected age and observable properties should be. During the next few years, the Tarter result was extrapolated too far forward in time and the Stevenson result was extrapolated too far backwards in time by theorists and observers eager to predict how many brown dwarfs might be detected with new space based instrumentation. The time dependence of (6) is so steep that the luminosity decreases rapidly and impedes the ability of the configuration to cool by radiation: these brown dwarf models maintain a higher T_e and thus produce objects better detected at optical wavelengths. The shallow time dependence of (1) leads to the more dramatic lowering of T_e and therefore favors detection in the infrared.

Using the Stevenson results, Staller and de Jong (1981) calculated that the Hubble Space Telescope Wide Field Camera should be able to detect 90 or so brown dwarfs per square degree to a limiting magnitude of 26, if the "missing mass" in the local solar neighborhood consisted of brown dwarf stars. Bahcall and Soneira (1980 a,b) published the first of what was to become a long series of papers by Bahcall attempting to model the various mass components of the Milky Way Galaxy, and any massive halo around it, in a manner that was self-consistent with the observational data on various stellar populations. In 1982, Mike Werner responded to Bahcall's growing interest in the possible role of brown dwarfs with a letter comparing the detection capabilities of SIRTF, HST, IRTF, the Luyten proper motion survey and IRAS. These calculations were performed by Ron Probst and were based on relation (1). Werner concluded that SIRTF could detect 100 brown dwarfs in 7 days of observing and further that the number of brown dwarfs detected as a function of galactic latitude could be used to distinguish between disk and halo populations.

During this time Abt and co-workers (Abt 1978, 1979) attempted to determine observationally the shape of the secondary mass function for multiple star systems. They concluded that more than 75% of all stars are members of multiple systems and that the short and long period binary systems have different secondary mass functions. Limits set by observational sensitivity were consistent with the secondary mass function of close, short period, binaries extending down to substellar companion masses. This reinforced the notion of a hierarchical fragmentation process that produced fragments below the hydrogen burning cut-off. This was counter to the conclusions of Scalo (1978) and Miller & Scalo (1979) who had deduced that the power law IMF, that well represented the upper portion of the main sequence, must turn down at masses above the hydrogen burning cut-off, thereby predicting few if any brown dwarfs and no significant contribution to the local "missing mass" problem.

"NEW BROWN DWARF PAPER"

The IRAS results quoted by Werner in the letter to Bahcall cited above are based on a paper by Reynolds, et al (1980). The performance characteristics of that satellite were well predicted, and its original operating schedule called for two complete surveys of the sky separated by 6 months. Reynolds et al argued that this was the ideal instrument to use in a search for either invisible solar system bodies (Bailey 1976) or "infrared dwarfs" (Davidson 1975) or a nearby companion star to the Sun (Harrison 1977 and Henricks & Staller 1978), all of which objects had been postulated in articles whose titles ended in question marks. The two surveys at a half year interval would provide the opportunity to use parallax and proper motions of these relatively nearby objects to distinguish them from the hundreds of thousands of other faint red point sources anticipated in the IRAS results. The two surveys could have been "blinked" against one another. Given the sensitivities of the four IRAS bands and the limited positional accuracy of the IRAS beam, we predicted that the reflected sunlight from a distant Neptune could be seen out to 10^3 AU and the internal energy sources of cool brown dwarfs could be detected and discriminated from other sources out to distances of 3×10^4 AU. For brown dwarfs hotter than 1000 K, we agreed with Staller and de Jong (1981) that the Hubble Space Telescope would be the instrument of choice.

"BROWN DWARFS 1981 + 1982 + 1983 +"

This folder contains a lot of notes in addition to published articles. Having made such bold predictions for the potential performance of IRAS, and being faced with the imminent launch of that spacecraft, we realsied that we really needed a more accurate prediction of the cooling curves for brown dwarfs than could be obtained from either of the Tarter or Stevenson power law extrapolations. Ray Reynolds therefore orchestrated a collaboration at NASA Ames Research Center with Martin Cohen, Al Grossman myself and some programmers. We knew that the opacities and the equations of state used in the old Grossman and Graboske attempts to determine the low mass end of the main sequence were highly suspect and in particular could not be expected to produce a believable color temperature for these objects. Another series of numerical models for the evolution of planetary mass objects had been developed at Ames (Pollack et al 1977) using similar, but different opacities and equations of state. Since we could not then (and still cannot now) specify what the correct opacities should be in evolving brown dwarfs, our plan was to evolve the stellar mass numerical codes to lower and lower masses and longer and longer times, while simultaneously evolving the planetary mass numerical codes to higher and higher masses and longer and longer times. In the region of mass and age overlap, we would have confidence in any gross parameters that were similarly predicted by both codes. At least we could argue that predictions arising from both sets of calculations were insensitive to the uncertain opacities and equations of state. This probably was a good plan, but the effort continued at a subcritical level and was frustrated by the introduction of several new computer systems over a few year period. This effort has not yet produced any results.

Although we did not succeed in producing a numerical code that could
properly be used to follow the evolution and cooling of brown dwarfs,
D'Antona & Mazzitelli (1982, 1983) did just that. In the process of
re-examining the value of the true low mass cut-off to the hydrogen
burning main sequence as a function of metallicity, these authors
investigated nuclear fusion reactions at the core of a configuration
that was not in thermal equilibrium due to a rapid mass loss. They
also pointed out the extreme precison necessary in determining the mass-
luminosity relationship for the low end of the main sequence and these
transitional non-equilibrium configurations before any conclusions can
be drawn about the inital mass function that produced them. Their predictions of luminosities, time scales and effective temperatures were
still subject to the intrinsic uncertainties in the opacities and
equations of state, but the models were carefully evolved forward in
time and did not have to rely on any power law extrapolations. Indeed,
had we been aware of these European efforts (that is, had we read A&A
carefully) we might have abandoned our sub-critical efforts altogether,
and this file folder would have contained fewer notes!

In the late 1970's Silk (1977 a,b,c) attempted to include evolving
metallicity and interactions (including coalescence) between fragments
in the analytic predictions of the minimum mass fragments and the initial mass function of star forming regions. In the early 1980's one,
two and three dimensional numerical calculations of "real" fragments
including rotation (Tohline 1981, Boss 1981, Bodenheimer 1981, Wood
1981) and magnetic fields (Scott & Black 1981, Mouschovias 1981) began
to appear. In addition, theorists (Silk & Norman 1980, Silk 1982,
Hunter & Fleck 1982) and numerical modellers (Woodward 1976, Sandford et
al 1982) began to consider the hierarchical fragmentation and formation
of protostars as taking place in the context of a dynamic and disturbed
environment due to the winds and shocks of nearby pre-main sequence
stars and supernovae explosions. As a result of all this activity, we
are probably closer now to a physically accurate description of the process of fragmentation, but unfortunately there is as yet no universally
accepted value for the minimum mass fragment that will form! During
this workshop, Alan Boss will present the latest calculations on the
fragmentation of rotating bodies undergoing collapse, and these calculations do indeed leave room for the formation of brown dwarf mass
fragments. However, a recent theoretical preprint from Larson (1985)
presents a far more pessimistic picture for those who desire a sufficient supply of brown dwarfs for local galactic gravitational stability and easy observational access.

During this time there continued to be observational searches for individual brown dwarfs as companions to faint low mass stars (Probst 1983)
and theoretical interest in constraining large ensembles of brown
dwarfs. Bahcall (1984) considered massive haloes around our galaxy and
Karimabadi & Blitz (1984) predicted the direct observational consequences of a closure density for the universe supplied by brown dwarfs.

"IRAS FOLLOW ON STUDIES"

IRAS flew an extremely successful mission, but unfortunately for our brown dwarf detection scheme, the two complete sky surveys separated by six months did not materialise. This file again contains a series of notes generated at meetings with Ray Reynolds, Russ Walker, Martin Cohen and myself at which we attempted to define sorting algorithms to extract brown dwarfs from the IRAS data base. A similar effort was being pursued by Frank Low of the IRAS Science Team, and he will report to you about his findings at this workshop. In our case we chose to work with the last operational database before the creation of the final IRAS point source catalog. Although the "months confirmation" that we desired had deteriorated to "weeks confirmation" over most of the sky, we still hoped to filter first on those sources that the IRAS software had found sufficiently discrepant to deny classification as a point source over the longest available time interval. To date Russ Walker has extracted hundreds of candidates and Martin Cohen has laboriously made visual identifications with Palomar plate objects: no brown dwarf detections! Thus far we have specifically avoided the galactic plane because of the enormous confusion of sources. In time that will get done, but the completeness limits that we will be able to place on this search are difficult to assess, depending heavily on the assumed orbit size and range of possible inclination angles attributed to each extracted candidate on the basis of its apparent magnitude.

"BROWN DWARFS FOR REAL!"

This of course is the really fun folder. The first entries here are due to Probst & Liebert (1983) and Liebert et al (1984) describing their studies of the strange object LHS 2924. They conclude that this is either "...(1) an old, low-mass star, probably defining a new empirical "bottom" to the main sequence, or (2) a substellar mass brown dwarf."

When the paleontological extinction event records and the impact cratering event records both seemed to predict the same surprising 26 to 28 million year periodicity, "Nemesis" was invoked as an explanation, and this folder gained a few more entries. Although Davis et al (1984) predicted that "Nemesis" should be an ordinary loss mass red dwarf, Whitmire & Jackson (1984) invoked a lower mass, less detectable brown dwarf. The validity of these periodicities deduced from small number statistics as well as the explanatory "Nemesis" hypothesis are still debatable. A positive detection by the Muller and his co-workers (Muller 1985) could settle the questions soon, and the observed orbital characteristics would then allow a mass inference to distinguish between the red and brown dwarf models.

The next item in this folder is the announcement of the discovery of VB8B by McCarthy, et al (1985); their inferred mass for the object making it a strong brown dwarf candidate! That was followed (with what seemed amazing rapidity, since I hadn't been keeping up with A&A) by a preprint from D'Antona & Mazzitelli (1985) that allowed for a far older

age to VB8B than had been deduced from the old brown dwarf cooling curves. Francesca D'Antona will describe her calculations and conclusions
for you later in this workshop, but the essence of the work is that if
you add grains or aerosols to the already uncertain opacities used in
the numerical codes, the resulting configuration is unable to cool very
rapidly and the central temperature remains sufficiently high for non-
equilibrium nuclear reactions to sputter along for a considerable length
of time, thus keeping the luminosity elevated. The VB8B announcement
and this workshop have produced a recent torrent of preprints (all of
which will be reported here at some time today). It's clearly time for
a new folder. I think I'll call it "DON'T LISTEN TO REFEREES!".

CONCLUSION
        Whether or not models for fragmentation during star formation
can predict them, we may have discovered at least one brown dwarf. The
following list of questions looms large at the start of this workshop,
and I suspect will continue to do so even at the end.

1. Have we indeed actually found any brown dwarfs? At the present
time, mass seems to be our sole criterion. This is insufficient because
it is never an unambiguously observed quantity. We need to be able to
predict color temperatures and broad spectral features, in order to
distinguish brown dwarfs from faint red stars having higher surface
gravities.

2. Do we know the correct equation of state? Perhaps near the asymptotic cooling-at-constant-radius part of the evolutionary track, when
these objects most resemble the gas giant planets, our equation of state
is adequate. However, for the regime where nuclear fusion has just
ceased to provide the luminosity, or does so in a non-equilibrium and
sporatic manner, all the details of non-zero temperature, screening
corrections, phase changes, electron conductivity and lack of thermal
equilibrium must be included in a more comprehensive way than has been
done to date. This is the phase of a brown dwarf's evolution where it
is most observable, whether it will be caught in this phase depends on
the time scale and recent work has cast much doubt on previous conclusions.

3. What is the correct form of atmospheric opacity to be used during
the different phases of brown dwarf evolution? This appears to be the
biggest single unknown and unfortunately the one to which numerical
models and observational predictions are the most sensitive.

4. Will a brown dwarf evolving in a multiple star system be strongly
influenced by "insolation" or gravitational effects?

5. Should we have found any brown dwarfs yet? What is the true low mass
stellar luminosity function and how does it convert to an initial mass
function? Can anything be extrapolated reliably beyond the current
observational magnitude limit?

6. What is the minimum mass fragment that forms and how does it differ among populations I, II and III?

7. Are current search techniques the most effective? Should we be trying to detect individual brown dwarfs or ensembles?

8. What about the name?

With respect to this last question, I have already explained how I originally adopted brown dwarf, and I have steadfastly maintained its usage during this paper. That is not to say that there have not been other names suggested over the years, some of them with great feeling. In Table 1. I have reproduced some of those suggested names and their sponsors, along with a brief annotation relating to what I feel are the

TABLE 1

Suggested name	By	Problem with the suggested name
Lilliputian Star	Shapley (1958)	Not stars: no equilibrium H burning.
Black Dwarf	Kumar (1963a)	Already means the end product of a chemically evolved White Dwarf.
Infrared Dwarf	Davidson (1975)	Too predictive: implies they are observable only in the IR whereas they may be young and hot or old and very cold.
"Jupiters" or Super-Jupiters	many authors	They do not necessarily require a star about which to orbit and may have formed in a fashion very different from the formation of the gas giant planets in a planetary system.
Extreme Red Dwarf	few authors	Admits a confusion about the end of the hydrogen burning main sequence rather than defining a difference.
Sub-stellar Objects	Reynolds et al (1980)	Can apply to planets as well.

difficulties with these other names. I am convinced that the non-color brown most aptly describes these bodies of questionable existence, whose formation probably has to do with a fragmentation process, and not the existence of a planetary system, whose opacities are unknown and whose appearance is therefore unpredictable. If the aerosol suggestion does turn out to be correct, then in the end a brown dwarf may even look brown, by which I mean similar to the colored pictures of Jupiter's atmosphere as rendered from the Voyager spacecraft data by the image reproduction wizards at JPL. As a last word, I think that it will be this workshop that will have settled the name question. If we were to change the name of the object, we would have to change the name of this workshop, and the title of its book of proceedings!

ACKNOWLEGEMENTS

This research was supported by NASA cooperative agreement NCC 2-36 between the NASA Ames Research Center and the University of California Berkeley, and NASA cooperative agreement NCC 2-336 to the SETI Institute.

REFERENCES

Abt, H.A. (1978) The Binary Frequency Along the Main Sequence in Protostars and Planets, pp 323-338, ed. T. Gehrels, University of Arizona Press.

Abt, H.A. (1979) The Frequency of Binaires on the Main Sequence. A.J. 84, 1591-1597.

Bahcall, J.N. (1984) Self-Consistent Determinations of the Total Amount of Matter near the Sun. AP.J. 276, 169-181.

Bahcall, J.N. & Soneira, R.M. (1980a) Star Counts as an Indicator of Galactic Structure and Quasar Evolution. Ap.J. (Letters) 238, L17-L20.

Bahcall, J.N. & Soneira, R.M. (1980b) The Universe at Faint Magnitudes I. Models for the Galaxy and Predicted Star Counts. Ap.J. Suppl. 44, 73-110.

Bailey, M.E. (1976) Can "Invisible" Bodies be Observed in the Solar System? Nature 259, 290-291.

Black, D.C. (1980) On the Detection of Other Planetary Systems: Detection of Intrinsic Thermal Radiation. Icarus 43, 293-301.

Bodenheimer, P. (1981) The Effects of Rotation During Star Formation in Fundamental Problems in the Theory of Stellar Evolution, pp 5-26. IAU Symposium 93, eds. D. Sugimoto, D.Q. Lamb, D.N. Schramm, D. Reidel Publ.

Boss, A.P. (1981) Collapse and Fragmentation of Rotating, Adiabatic Clouds. AP.J. 250, 636-644.

Cox, A.N. & Stewart, J.N. (1970) Rosseland Opacity Tables for Population I Compositions. Ap.J. Suppl. 19, 243-279.

D'Antona, F. & Mazzitelli, I. (1982) Evolution of Low Mass Stars Through Mass Loss: Transition from the Main Sequence to the Degenerate Phase. Astron. and Astrophys. 113, 303-310.

D'Antona, F. & Mazzitelli, I. (1983) Mass Luminosity Relation and Initial Mass Function at the Faint End of the Main Sequence. Is there a Real Deficit of Low-Mass Stars? Astron. & Astrophys. 127, 149-152.

D'Antona, F. & Mazzitelli, I. (1985) Evolution of Very Low Mass Stars and Brown Dwarfs I. Minimum Main Sequence Mass and Luminosity. AP.J. 296, 502-513.

Davidson, K. (1975) Does the Solar System Include Distant but Discoverable Infrared Dwarfs? Icarus 26, 99-101.

Davis, M., Hut, P. & Muller, R.A. (1984) Extinction of Species by Periodic Comet Showers. Nature 308, 715-717.

Field, G.B. (1970) in Evolution Stellaire Avant la Sequence Principale pp 49. 16th Liege Symposium.

Gaustad, J.E. (1963) The Opacity of Diffuse Cosmic Matter and the Early Stages of Star Formation. AP.J. 138, 1050-1073.

Grossman, A.S. (1970) Evolution of Low-Mass Stars I. Contraction to the Main Sequence. AP.J. 161, 619-632.

Grossman, A.S., Mutschlecner, J.P. & Pauls, T.A. (1970) Evolution of Low-Mass Stars II. Effects of Primoridal Deuterium Burning and Nongray Surface Condition During Pre-Main-Sequence Contraction. AP.J. 162, 613-619.

Grossman, A.S. & Graboske, H.C. Jr. (1971) Evolution of Low-Mass Stars III. Effects of Nonideal Thermodynamic Properties During the Pre-Main-Sequence Contraction. AP.J. 164, 475-490.

Grossman, A.S., Hays, D. & Graboske, H.C. Jr. (1974) The Theoretical Low Mass Main Sequence. Astron. & Astrophys. 30, 95-103.

Harrison, E.R. (1977) Has the Sun a Companion Star? Nature 270, 324-326.

Heinricks, H.F. & Staller, R.F.A. (1978) Has the Sun Really Got a Companion Star? Nature 273, 132-134.

Heintz, W.D. (1972) Astrometric Study of Four Binary Stars. A.J. 77, 160-165.

Hubbard, W.B. (1968) Thermal Structure of Jupiter. AP.J. 152, 745-754.

Hubbard, W.B. & Slattery, W.L. (1971) Statistical Mechanics of Light
        Elements at High Pressure I. Theory and Results for Metallic
        Hydrogen with Simple Screening. AP.J. 168, 131-139.

Hunter, J.H. Jr. & Fleck, R.C.Jr. (1982) Star Formation: The Influence of
        Velocity Fields and Turbelence. AP.J. 256, 505-513.

Hyashi, C. (1962) Publ. A.S.J. 13, 450.

Karimabadi, H. & Blitz, L. (1984) The Detectability of Population III
        "Jupiters". AP.J. 283, 169-173.

Kumar, S.S. (1963a) The Structure of Stars of Very Low Mass. AP.J. 137,
        1121-1125.

Kumar, S.S. (1963b) The Helmholtz-Kelvin Time Scale for Stars of Very
        Low Mass. AP.J. 137, 1126-1128.

Kumar, S.S. (1972) Hidden Mass in the Solar Neighborhood. Ap. and Sp.
        Sci. 17, 219-222.

Larson, R.B. (1973) A Simple Probabilistic Theory of Fragmentation.
        MNRAS 161, 133-143.

Larson, R.B. (1985) Bimodal Star Formation and Remnant-Dominated
        Galactic Models. Preprint to be published in MNRAS.

Liebert, J., Boroson, T.A. & Giampapa, M.S. (1984) New Spectrophotometry
        of the Extremely Cool Proper Motion Star LHS 2924. AP.J.
        282, 758-762.

Lipincott, S.L. & Hershey, J.L. (1972) Orbit Mass Ratio and Parallax of
        Visual Binary Ross 614. A.J. 77, 679-683.

Luyten, W.J. (1968) A New Determination of the Luminosity Function.
        MNRAS 139, 221-224.

Lynden-Bell, D. (1973) in Dynamical Structure and Evolution of Stellar
        Systems. Sauverny, Switzerland: Geneva Observatory.

McCarthy, D.W. Jr., Probst, R.G. & Low, F.J. (1985) Infrared Detection
        of a Close Cool Companion to Van Biesbroeck 8. AP.J.
        (Letters) 290, L9-L13.

Miller, G.E. & Scalo, J.M. (1979) The Initial Mass Function and Stellar
        Birthrate in the Solar Neighborhood. AP.J. Suppl. 41,
        513-547.

Mouschovias, T.C. (1981) The Role of Magnetic Fields in the Formation of Stars in Fundamental Problems in the Theory of Stellar Evolution, pp 27-62. IAU Symposium 93, eds. D. Sugimoto, D.Q. Lamb & D.N. Schramm, D. Reidel Publ.

Muller, R.A. (1985) Evidence for a Solar Companion Star in The Search for Extraterrestrial Life: Recent Developments, pp 233-243, ed. M.D. Papagiannis, D. Reidel Publ.

Ostriker, J.P. & Axel, L. (1969) On the Cooling of White Dwarfs in Low Luminosity Stars, ed. Kumar pp 357-363, Gordon and Breach Publ.

Peebles, P.J.E. (1964) The Structure and Composition of Jupiter and Saturn. AP.J. 140, 328-347.

Pollack, J.B., Grossman, A.S., Moore, R. & Graboske, H.C. Jr. (1977) A Calculation of Saturn's Gravitational Contraction History. Icarus 30, 111-128.

Probst, R.G. (1983) An Infrared Search for Very Low Mass Stars: JHK Photometry and Results for Composite Systems. AP.J. Suppl. 53, 335-349.

Probst, R.G. & Liebert, J. (1983) LHS 2924: A Uniquely Cool Low-Luminosity Star with a Peculiar Energy Distribution. AP.J. 274, 245-251.

Reynolds, R.T., Tarter, J.C. & Walker, R.G. (1980) A Proposed Search of the Solar Neighborhood for Substellar Objects. Icarus 44, 772-779.

Salpeter, E.E. (1955) The Luminosity Function and Stellar Evolution. AP.J. 121, 161-167.

Salpeter, E.E. (1973) On Convection and Gravitational Layering in Jupiter and in Stars of Low Mass. AP.J. (Letters) 181, L83-L86.

Salpeter, E.E. & Zapolsky, H.S. (1967) Phys. Rev. 158, 876.

Sandford, M.T. II, Whitaker, R.W. & Klein, R.I. (1982) Radiation Driven Implosion in Molecular Clouds. AP.J. 260, 183-201.

Scalo, J.M. (1978) The Stellar Mass Function in Protostars and Planets, pp 265-287, ed. T. Gehrels, University of Arizona Press.

Scott, E.H. & Black, D.C. (1981) The Role of Magnetic Fields in the Collapse of Protostellar Gas Clouds in Giant Molecular Clouds in the Galaxy, pp 303-311, eds. P.M. Solomon & M.G. Edmunds, Pergamon Press.

Shapley, H. (1958) in Of Stars and Men, p 60, Beacon Press, Boston.

Silk, J. (1977a) On the Fragmentation of Cosmic Gas Clouds I. The
        Formation of Galaxies and the First Generation of Stars.
        AP.J. 211, 638-648.

Silk, J. (1977b) On the Fragmentation of Cosmic Gas Clouds II.
        Opacity-Limited Star Formation. AP.J. 214, 152-160.

Silk, J. (1977c) On the Fragmentation of Cosmic Gas Clouds III. The
        Initial Stellar Mass Function. AP.J. 214, 718-724.

Silk, J. (1982) Does Fragmentation Occur on Protostellar Mass Scales
        During the Dynamic Collapse Phase? AP.J. 256, 514-522.

Silk, J. & Norman, C. (1980) Clumpy Molecular Clouds: A Dynamic Model
        Self Consistently Regulated by T Tauri Star Formation. AP.J.
        238, 158-174.

Staller, R.F.A. & de Jong, T. (1981) Theoretical Luminosity Function of
        Red and Black Dwarfs. Astron. & Astrophys. 98, 140-148.

Stevenson, D.J. (1978) Brown and Black Dwarfs: Their Structure,
        Evolution and Contribution to the Missing Mass. Proc.
        Astron. Soc. Australia 3, 227-229.

Stevenson, D.J. (1979) Solubility of Helium in Metallic Hydrogen.
        J. Phys. F. 9, 791-801.

Stevenson, D.J. & Salpeter, E.E. (1977) The Dynamics and Helium
        Distribution in Hydrogen-Helium Fluid Planets. AP.J. Suppl.
        35, 239-261.

Tarter, J.C. (1975) Brown Dwarf Stars and How Long They Grew Old. in
        unpublished Ph.D. Thesis, University of California, Berkeley.

Tinsley, B.M. (1973) Analytic Approximations to the Evolution of
        Galaxies. AP.J. 186, 35-49.

Tohline, J.E. (1981) The Collapse of Equilibrium Rotating Adiabatic
        Spheroids I. Protostars. A.J. 248, 717-726.

Veeder, G.J. (1974) Luminosities and Temperatures of M Dwarf Stars from
        Infrared Photometry. A.J. 79, 1056-1072.

Weistrop, D. (1972) The Luminosity Function and Density Distribution of
        Disk Population Stars. A.J. 77, 849-862.

Werner, M. (1982) private communication, letter to John Bahcall.

Whitmire, D.P. & Jackson, A.A. IV (1984) Are Periodic Mass Extinctions
        Driven by a Distant Solar Companion? Nature 308, 713-715.

Wood, D. (1981) Collapse and Fragmentation of Isothermal Gas Clouds.
        MNRAS 194, 201-218.

Woodward, P.R. (1976) Shock Driven Implosion of Gas Cloud and Star
        Formation. AP.J. 207, 484-501.

QUESTIONS AND ANSWERS
        D'Antona: What are the typical densities you used to derive
the Debye temperature and what values for the Debye temperature did you
get?

Tarter: The following are the values I calculated for several different
mass configurations:

Mass	Central density of zero temperature configuration: ρ_0	Debye Temperature: $T_D(\rho_0)$
.08 M_o	2716 gm cm^{-3}	1.7 x 10^5 K
.03 M_o	537 gm cm^{-3}	7.4 x 104 K
.01 M_o	105 gm cm^{-3}	3.3 x 10^4 K
.003 M_o	25 gm cm^{-3}	1.6 x 10^4 K

Zinnecker: Suppose the number density of brown dwarfs would be as high
as to provide the closure mass density of the universe, would their
integrated background light in the far infrared be detectable (e.g. with
SIRTF)?

Tarter: Karimabadi & Blitz (1984) have calculated that a detectable
effect might exist in the 10-100 µm wavelength range.

SIGNIFICANCE OF BROWN DWARFS

D. C. Black, NASA Ames Research Center and
            NASA Headquarters

Abstract. Scientific interest in objects that
are not massive enough to qualify as stars has,
until recently, not been high. However, the
advent of powerful and clever new observational
techniques, as well as of new instrumentation
such as the Hubble Space Telescope, signals the
dawn of a new era in research related to
sub-stellar mass bodies.

The significance of these bodies in terms of
their role in furthering our understanding of
major problems in astronomy is discussed from an
admittedly subjective point of view. The
discussion centers on two themes; brown dwarfs
in their own right, and brown dwarfs and their
relationship to planetary systems.

1   INTRODUCTION

When I was asked several months before the
Symposium if I would speak on the "Significance of Brown
Dwarfs", I did what many of us do--I accepted without
really thinking about what I had agreed to discuss. As I
prepared for this lecture it became clear that these
objects could potentially hold keys to many of the more
challenging problems in astronomy and astrophysics.

It also became clear that there are nearly as many notions
of what a brown dwarf is as there are papers on the
subject. Therefore, in the interest of clarity I am
compelled to offer here the definition of "brown dwarf"
that I shall use throughout this discussion. I make no
claim that this definition is fundamentally better than
others, but it does seem to define the class in terms of
those things that as scientists we should focus our
attention on, physical processes. I define a brown dwarf
as follows:

> "A brown dwarf is any sub-stellar
> mass body formed by the same process
> that forms stars."

This definition provides an upper limit to the mass of a
brown dwarf, viz., that required to burn hydrogen.
Quantitatively, this limit is about 0.075 $M_\odot$, with an
uncertainty of 0.005 $M_\odot$. It has been well documented by a
variety of workers that nuclear burning of deuterium can
occur in objects of mass as low as 0.015 $M_\odot$, but as the
duration of that phase is short compared to other phases
in the life of these bodies (in contrast to main-sequence),
I do not consider that significant for the purpose of this
paper.

This definition does not establish explicitly a lower
limit to the mass of a brown dwarf, however, such a limit
is prescribed implicitly in the requirement that one is
concerned with objects that are formed by the same process
or processes that form stars (see paper by Boss in these
proceedings). This approach may seem to be somewhat
academic because we do not know in detail how stars form.
But it is just this focus on process which I feel is
crucial in understanding the significance of brown dwarfs.
In some sense then brown dwarfs are what McNally has
referred to as "failed stars", although he was applying
that description, incorrectly in my view, to Jupiter.

This discussion is presented in two main parts; one
dealing with areas of physics and/or astrophysics in which
increased knowledge of brown dwarfs can play a major role,
and one dealing with an area of research in which the role
of brown dwarfs is less obvious, but which nonetheless has
received the most attention. This latter area is that of
other planetary systems.

2   BROWN DWARFS AND ASTROPHYSICS
One can identify several areas of research in
physics and astrophysics in which knowledge gained from
studies of brown dwarfs could have significance. I will
not attempt to give a complete discussion of these as many
of them are covered in detail elsewhere in these proceedings. I will try to indicate those which I think are
of particular significance, and why that is so.

One of the more important areas in which knowledge of
brown dwarfs will play a major role is that of star
formation. This role comes both from brown dwarfs as
isolated bodies and as members of binary systems.

At the present time we have a reasonably good working
model of at least the general phases and processes that
are involved in the formation of stars. For example, it
is becoming clear that most stars begin their protostellar
evolution in dense regions, often referred to as
"fragments", of molecular clouds known as molecular cloud
cores. Protostellar evolution involves a complex
interaction between a variety of effects, including
rotation, magnetic fields, gravity, and some form of what
is usually called "turbulence" (the quotes are given
because this protostellar "turbulence" is not that found
in laboratory experiments). The outcome of this
interaction in most cases is another stage of
fragmentation, one that generally produces a binary
system.

Our understanding of what controls this latter fragmentation process, and this is the one that establishes the
mass function for stars, is incomplete. An observational
determination of the mass function of brown dwarfs will
greatly extend and constrain our models of this important
process. Studies of brown dwarfs in binary systems are
particularly relevant here as that is the only way to
obtain an independent measure of the mass of a brown
dwarf. However, one must be cautious. Studies by Abt
(1978) have shown that the mass function for companions in
short period binaries, those with orbital periods of 100
years or less, differs from those of longer period
binaries. The latter mass function appears to match that
of field stars. Thus, as most of our information on
masses of brown dwarfs is likely to come from shorter
period binary systems, it is not clear that that data
pertains to the mass function of brown dwarfs in the
field. That in no way diminishes the importance of such
observations.

One other aspect of brown dwarfs which has potential
significance involves time, or more correctly age.
Although researchers have been able to date both in a
relative and an absolute sense samples of extraterrestrial
material, it is very difficult to obtain accurate estimates of the age of astronomical systems. If we are able
to develop good theoretical models of brown dwarfs, models
which permit detailed comparison of observable properties
with the age of a brown dwarf, it may be possible to
obtain independent measures of the ages of stars in binary
systems. A clock of that type would be of great value to
stellar evolution theory.

There are other areas where studies of brown dwarfs are
certainly of value but not, in my view, of the same

significance as the above examples. Brown dwarfs are
degenerate objects, bodies in which the equation of state
that describes the behavior of material is non-ideal and
involve a variety of interesting quantum effects. The
potential of these bodies as celestial laboratories in
which nature provides an environment for study of this
unusual state of matter, is often overstated. Much of our
current understanding of such systems comes from
observations of the giant planets in the solar system,
bodies that can be studied in detail by means of
spacecraft, permitting formulation and testing of complex
theoretical models of mildly degenerate bodies. It is
unlikely that we will ever be able to study brown dwarfs
in the detail that we can study degenerate planets in our
own planetary system. In fact, it is reasonably certain
that knowledge of the equation of state gained from
studies of the giant planets is adequate for most analyses
of the behavior of brown dwarfs.

An important exception to this concerns the equation of
state for bodies right at the transition mass between a
star and a brown dwarf. It is important to refine our
understanding here in order to construct better models of
the interior nuclear processes and thereby of the
observable signature of these transition objects. Also,
as brown dwarfs are only mildly degenerate they do not
provide insight into the intricacies of the world of
particle physics. One needs strongly degenerate systems
such as neutron stars to examine the basic behavior of
neutronic fluids.

One other area which is often mentioned with regard to
brown dwarfs is that of galactic dynamics. This is often
referred to as the "missing mass" problem in galactic
astronomy. This terminology is unfortunate--the mass is
not "missing", we simply do not know what form the mass is
in. That it exists is clear from the dynamic behavior of
the visible components of the galaxy in the solar neighborhood. If it should turn out that observational studies
show that the mass function of brown dwarfs is adequate to
account for the dynamically required mass, it will help
resolve a longstanding problem in galactic astronomy.

3 BROWN DWARFS AND OTHER PLANETARY SYSTEMS
The study of brown dwarfs has received increased
attention recently because of the possibility that these
sub-stellar objects are in some way related to planets.
In fact, much of the impetus in development of new obser-

vational techniques has come from a desire to detect other planetary systems. A notable example of this interest can be found in the observation by McCarthy et al. (1984) of the companion to the low-mass star Van Biesbroeck 8. The key question to be addressed here is "What is the relationship between brown dwarfs and planets?"

The reader will note that I said searching for other planetary systems, not searching for extrasolar planets. The distinction is deliberate, and relates to my introductory remarks concerning the need for scientists to focus on processes. We search for other planetary systems because it is essential to understand how the solar system formed. The emphasis is on a system and its formation, not just a single body such as a companion to another star. An extensive discussion of the issues and problems associated with detection of other planetary systems is given elsewhere (Black & Levy 1985). I will summarize that discussion here.

Somewhat surprisingly there is no generally agreed upon definition of what constitutes a "planet". A planet is a sub-stellar mass object, and so has something in common with a brown dwarf. However, there are a variety of reasons to believe that planetary systems, and planets, are formed by a different process than that which forms stars. The highly structured character of the solar system suggests that it was formed from a highly dissipative structure, and that the process of formation was one which did not give rise to bodies of cosmic composition. Even the largest planet in the solar system is not of solar composition; it has in comparison with the Sun an overabundance of heavy elements relative to hydrogen. This fact, and the fact that most of the outer planets appear to have similar "rocky" cores, suggests that they were formed by a process of accretion from small objects. This is fundamentally different from the bulk hydrodynamic process that is believed to be the manner in which stars, and hence brown dwarfs, form.

This compositional signature is a small effect for large planets, and prior to attending the symposium I would have guessed that it could not be of use in discriminating sub-stellar bodies formed by accretion from those formed by large-scale gravitational collapse. However, in light of the paper given by Lunine et al. at this symposium, that conclusion needs to be re-examined critically.

One might ask why detection of a single sub-stellar mass companion to a star is not adequate evidence of detection

of another planetary system. It is important to recognize
that nature prefers to make stars in binary systems.
Current surveys suggest that at least 70 percent of all
stars are in binary systems, and this number must be
considered as a lower limit. There is no reason to
suppose that the binary formation process will only make
stellar binary systems. The process of making a binary
does not involve nuclear physics, and so it is entirely
reasonable that binary systems can be formed where one
member is a star and the other is too small to burn
hydrogen. For this reason, it is generally felt that one
must find a star with more than one sub-stellar companion
to be confident that a planetary system has been
discovered.

Thus, although statistical information on brown dwarf
companions to nearby stars is of great interest and value
to a number of research areas in astrophysics, such data
cannot be directly taken to say anything about the nature
or even the existence of other planetary systems.

4 CONCLUDING REMARKS

I have tried to give some indication of the wide
range of problems in which observational and theoretical
studies of brown dwarfs will significantly advance our
understanding. I have also indicated problems where I
think that the significance of brown dwarfs is less
important, or perhaps of only marginal relevance.

Questions of significance aside, there is no question that
we are witnessing the dawn of a new era in astronomical
research, one in which new observational techniques and
instrumentation will permit us to detect and study a class
of objects that will pose new challenges to astrophysics
on the one hand, and answer long-standing questions on the
other.

References

Abt, H. A. (1978). The binary frequency along the main
       sequence. In Protostars and Planets, ed.
       T. Gehiels, pp. 323-37. Tucson: University of
       Arizona Press.

Black, D. C. & Levy, E. H. (1986). What is a planetary
       system? Submitted to Nature.

Lunine, J. I., Hubbard, W. B., and Marley, M. S. (1986).
       Atmospheres of Brown Dwarfs: A comparison with
       jovian planets. In Astrophysics of Brown
       Dwarfs, eds. R. Harrington & M. Kafatos.

Discussion

W. B. Hubbard: There is a well-known progression of
hydrogen depletion as we go down in mass from Jupiter
to Saturn to Uranus/Neptune. Extrapolating this trend
in the direction of increasing mass we would expect
objects to become more solar in composition. Does this
argue against the sharp dichotomy in formation
mechanisms which you propose?

D. C. Black: The short answer to your question is "no",
but let me go further. The essential part of my
proposed dichotomy is that there is a maximum mass
for a secondary growing by means of accretion,
actually mass ratio between the accreting body and
that of the central object (star), at which further
accretion is either terminated or significantly
retarded. This "truncation" could be due to tidal
interactions, as have been suggested by a number of
authors, to magnetic fields--no adequate theoretical
models yet exist. More massive secondaries must then be
formed by a global process. A consequence of this view
is that when truncation occurs it permits, indeed may
foster, accretion around seed bodies in other orbits.
In contrast, the global process could well preclude
formation of such companions. Note that in this view it
is perfectly possible to have "brown dwarf" (i.e.,
binary) companions to certain types of stars which are
less massive than "planetary companions" to other types
of stars! For example, if truncation occurs at a fixed
mass ratio, say 10^{-2}, then a brown dwarf companion to an
$\mathcal{M}$ dwarf could have a mass as low as a few Jupiter
masses whereas we may find planetary systems associated
with F dwarfs where the most massive companion has a
mass of a few tens of Jupiter masses. I remind you that
this concept derives from my best guess based on a
decade of theoretical work on star and planetary system
formation. It does not come from any specific
calculation.

J. Liebert: Is there a maximum mass ratio of planetary
disk to central star above which the formation of
planets by the core accretion process is impossible?
Presumably this might translate into a lower limit
for the mass ratio, M_1/M_2, of a binary or multiple
star system.

D. C. Black: We do not know the answer to your
first question, but I suspect that it is not the mass
of the disk that is important in setting the upper
limit to growth by accretion. Whatever controls
truncation of accretional growth is likely to be
relatively local parameter, not a global parameter

such as disk mass. Several studies have indicated that
there is a <u>minimum</u> ratio of disk to central star mass
for onset of <u>global</u> instabilities; this minimum value is
∽1/3 with considerable uncertainty. Whether this sets a
lower limit to a binary mass ratio is unclear. Alan
Boss has argued in his paper given at this symposium
that the lowest mass that can be formed by a
star-formation like process is $\simeq 0.02$ $M_\odot$, more or less
independent of the mass of the central star.

R. Probst: I would like to comment that the studies
by Abt and coworkers on the equivalence of binary
companion and field star mass functions is only on
companions to F-G main sequence stars and only down
to masses $\sim 0.4 M_\odot$. The assumption, which I and others
make, that this equivalence also holds for brown
dwarf companions to much different primaries, is a
leap of faith. Better binary surveys, i.e., more
complete and with better definition, among the Gliese
stars using the Hubble Space Telescope could serve to
test this leap.

D. C. Black: An excellent point. This is why I think
that the work that you, Don McCarthy, and others are
doing is so important.

A. P. Boss: Given that there are two mechanisms, one
for making brown dwarfs with a minimum mass of about
$0.02 M_\odot$, and another for making planets like Jupiter,
how large could a "Jupiter" be? Is there a gap
between the maximum planet mass and the minimum brown
dwarf mass?

D. C. Black: As you are aware, the theory of planetary
system formation is not yet sufficiently developed to
be able to yield reliable predictions of the maximum
mass of a planet as a function of, say, stellar mass.
I would predict that the maximum mass does vary with
spectral type of the parent star, but neither the
magnitude <u>nor</u> the sense of such a variation is known.
My guess is that if one concentrates on stars of a
<u>given</u> spectral type, then a gap of the type you ask
about will exist. However, as I indicated in my
response to Bill Hubbard's question, there is no <u>a</u>
<u>priori</u> reason to expect that such a gap exists when
comparing stars of <u>different</u> spectral types.

N. Roman: What is the definition of a brown dwarf
as distinguished from a star? Can it have burned D,
Li, Be, etc.?

D. C. Black: The definition that I offer here
distinguishes a star from a brown dwarf solely on the
basis of mass. A star is massive enough to sustain
itself by converting hydrogen into helium, thus it must
be more massive than $\simeq 0.075 M_\odot$. A brown dwarf is any
sub-stellar mass object that is formed by the same
processes that form stars. Objects of mass $\gtrsim 0.015 M_\odot$ are
capable of burning deuterium, so any brown dwarf in the
mass range $0.015 \lesssim M \lesssim 0.075 M_\odot$ will experience limited
nuclear burning. Whether nature forms brown dwarfs with
masses $< 0.015 M_\odot$ is an open question, but such objects
would not burn even deuterium.

H. Zinnecker: Does the formation of a disk surrounding
a central condensation imply the formation of a
planetary system rather than a binary star?

D. C. Black: No. The controlling factor is whether
companion formation is by means of an accretional or
a bulk hydro (perhaps magneto hydro) dynamic process.
That will depend on details of the disk's properties
and evolution. It may well be that both processes
can be operative in some systems. That is, planetary
systems may form in association with binary systems.
There is circumstantial evidence based on orbital
geometries that planetary systems form out of highly
dissipative disks (regularized orbits) whereas
binaries form out of less dissipative disks (I am not
considering wide binaries which appear to have formed
as independent systems). This may reflect a temporal
sequence; binaries form early in disk evolution and
planetary systems late in disk evolution, or it may
reflect a basic difference in the types of disks
which form binaries from those which form planetary
systems. Much more work, both observational and
theoretical, is required in this area.

VERY LOW MASS STARS VERSUS BROWN DWARFS:
A COMMON APPROACH

Francesca D'Antona
Osservatorio Astronomico di Roma, I-00040 MONTE PORZIO

Abstract. A unique input physics and evolutionary approach
is adopted to build evolutionary models of very low mass
stars and brown dwarfs. The model results lead to define a
boundary class, which will be called 'the transition
masses', whose constituents are objects which live about
10 billion years at luminosities between 10^{-4} and 10^{-5} Lo,
and are partially supported by nuclear burning and
partially by thermal cooling. Reasons are given to support
the suggestion that VB8B is a transition mass.
The special importance of using the appropriate mass
luminosity relation, to derive correctly the initial mass
function of very low mass stars, is stressed, and it is
predicted that there should be about 6 brown dwarfs in a
radius of 0.5 pc from the Sun, if they have to account for
dark matter density of 0.1Mo pc^{-3} in the solar neighborhood.

1. INTRODUCTION

In this review I discuss a number of subjects which are of interest in
the context of the astrophysics of Brown Dwarfs (BDs), and which can be
addressed from the point of view that there is a structural continuity
between stars and BDs. For a more complete presentation of the subject,
I refer to a series of papers on the evolution of Very Low Mass stars
(VLMs) by D'Antona and Mazzitelli (1982, 1983, 1985 and 1986, here-
inafter referred to as DM plus the year of publication).

First of all I will give the 'definitions' with the help of
our model results: what distinguishes a VLM star from a BD? This will
be useful, later on, to understand the evolutionary meaning of the
first 'bona fide' BD observed, VB8B (McCarthy et al 1985). Then I will
discuss the uncertainties inherent in the structures computation, what
we need to know before we may produce better models, and what conclusions can however be drawn on the basis of present generation models.

As it can be of interest for observers working on
deep surveys stellar counts (e.g. Boeshar and Tyson 1985,
Boeshar et al. 1985, Hawkins 1985) I will concentrate on which
predictions come out for the stellar luminosity function from the
assumption of a Salpeter type Initial Mass Function (IMF) for the low

mass end of the main sequence. I will stress that no reasonable predictions can be done if we do not make use of the proper mass-luminosity
relation for VLMs, which is now well established from theory.

Finally, I will examine the informations on the age of
the system containing VB8B, and conclude that there is no convincing
indication pointing to a young age. A lower limit of 2×10^9 yr is to
be assigned to the age of VB8b, which consequently is of mass very
close to the hydrogen burning minimum mass, and is at least partially
supported by nuclear burning.

Figures 1 & 2: Central temperature and luminosity versus
age for VLMs and BDs

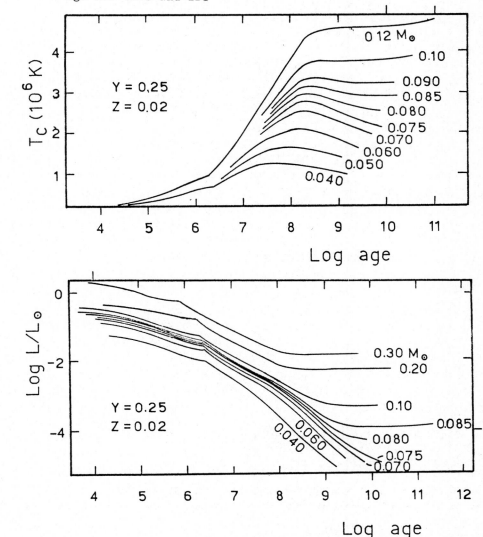

2. DEFINITIONS AND UNCERTAINTIES

Figures 1 and 2 will help me to introduce the 'definitions', as they show the run with time of central temperature (Tc) and total luminosity for VLMs and BDs. The data refer to the results by DM 1985, where complete informations on the model computations and input physics are given. Here I must only remember that we have taken care of adopting a good equation of state (e.o.s.), adequate to describe correctly the complex physics of the partial dissociation and ionization regions (Magni and Mazzitelli 1979). Obviously, also all the important corrections to the e.o.s. in full ionization (degeneracy, coulomb and exchange effects) are taken into account, as well as the appropriate weak and intermediate screening corrections to the nuclear reaction rates (Graboske et al. 1973).

The models are initially at large radius and low central temperature ($Tc=10^5$ K). In the first gravitational contraction phase, the radius decreases and Tc and central density increase. The figures show that Tc and total luminosity have a short stop, lasting for about 10^6 yr, due to the burning of primeval deuterium (2.5×10^{-5} in mass fraction). After that, the contraction (which is never completely halted) proceeds.

For relatively large masses ($M > 0.08 M_\odot$), finally the nuclear energy is able to support completely the luminosity, and the structure reaches a stable configuration: we have a VLMs, which will live for longer than a Hubble time in the phase of proton-proton burning.

For very small masses, in the range from 0.04 to $0.06 M_\odot$, the rise in Tc is halted by the onset of degeneracy before there is any significant release of nuclear energy: a phase of 'cooling' ensues, during which the source of luminosity is mainly the release of the thermal content of the ions, with a decrease in Tc. This phase lasts up to 5×10^9 yr down to $10^{-5} L_\odot$. These structures are defined BDs, or degenerate hydrogen dwarfs, to distinguish them from white dwarfs.

The definition of Hydrogen Burning Minimum Mass (HBMM) comes out immediately: it is the minimum mass for which stabilization of the structure in main sequence is possible. This mass (and, as we will see, mainly the minimum main sequence luminosity), depends (once fixed the helium content, see DM 1982) on the opacity behaviour in the atmospheric layers, so that there are important differences between the population I and population II low end of the main sequence.

Between the HBMM and the realm of BDs there exists a small range of masses whose structures behave intermediately: after a short phase (about 10^8 yr) of gravitational contraction, the central conditions allow a significant nuclear energy release, although it is insufficient to stabilize the structure, which also cools. An interesting finding of DM 1985 is that these structures may be very long lived, they spend a time comparable to the disk lifetime at luminosities between 10^{-4} and 10^{-5} L⊙, and I will call them 'transition masses'. It is almost obvious that transition masses must exist, for any choice of external opacities, but they are much more long lived when the atmospheric opacities are large.

This brings us to the major point of uncertainty: the computation of opacities. The opacity tables used in DM 1985 are taken from Alexander et al. (1983). These constitute an improvement with respect to Cox and Stewart (1970) tables, as the main molecules are taken into account at T<4000K. As shown in DM 1985, the major effect of inclusion of molecules in the opacities is to reduce the model luminosity. This consequence was already known, and it was even more pronounced with the use of the previous opacities by Alexander (1975), in which the influence of molecules was largely overestimated. In fact, the models computed by Alexander (1975) tables were so much underluminous to be in contrast with the main sequence observational mass-luminosity (ML) relation (e.g. Sienkiewvicz 1982, DM 1985). On the contrary, the agreement between our models and the observational ML relation can be taken as an indication that these new molecular opacities are reasonably correct for VLMs.

Smaller model luminosity for given central conditions has two main effects:

i) stable main sequence models are found down to luminosities much smaller than previously given. Our 0.085M⊙ main sequence model is at logL/L⊙=-3.95 while the equivalent by Grossman et al 1974 is at logL/L⊙=-3.30. More uncertain are the models of smaller mass, as they evolve at Teff<2000K, where the grain opacities, which are included in Alexander et al (1983) tables begin to be important. Unfortunately, these tables are not extended enough towards the large densities characteristic of VLMs in the atmosphere, at these temperatures, so that we had to extrapolate opacities both on temperature and density. Taken with caution, we find that also the 0.08Mo reaches the main sequence, at logL/L⊙=-4.35, after a pre-main sequence phase lasting about 10^{10} yr.

ii) During the evolution, the object 'understands its destiny' (whether the nuclear output will be sufficient of not to stabilize the structure) at smaller luminosities than with the use of Cox-type opacities. Therefore, the transition masses are very long lived, as they release their thermal energy content at very low luminosities. Also using Cox-type opacities transition masses are found (see figure 8 in DM 1985), but their luminosities are much larger, for a given Tc, so that the cooling lasts for a shorter time. A similar situation is found when considering the cooling of carbon-oxygen white dwarfs: the smaller is the atmospheric opacity, the faster the cooling (Mazzitelli and D'Antona 1986). Therefore, although wider grids of grain opacities are needed before we may build up more realistic models of transition masses, their temporal evolution is in any case dependent not on the actual grain opacity at $L=10^{-4}$ - 10^{-5} L$\odot$, but mostly on the previous opacity behaviour (mainly molecular), which has determined the thermal structure of the star at larger luminosity. As a conclusion, if Alexander et al. (1983) molecular (not grain) opacities are reasonably correct, transition masses are very long lived (10 -13 billion yr down to 10^{-5} Lo). If Cox-type opacities apply, transition masses still exist, but they live at most 5-6 billion years.

3. THE INITIAL MASS FUNCTION FOR VLMS

A result which arises from all recent studies of VLM main sequence structures is that the ML relation can not be expressed by a power law (a linear logM-logL/L$\odot$ relation) at M<0.2M$\odot$. The reasons for this behaviour are discussed in DM 1982, and can be summarized as due to two combined effects: lowering of the adiabatic gradient in the subatmosphere, when the models cross the molecular hydrogen dissociation region, and very rapid decrease of the nuclear energy release by p-p burning by small variations of central conditions. The consequences can

Figure 3: dots: MF of nearby stars. Histogram: LF of nearby stars transformed into the mass intervals corresponding to one visual magnitude intervals.

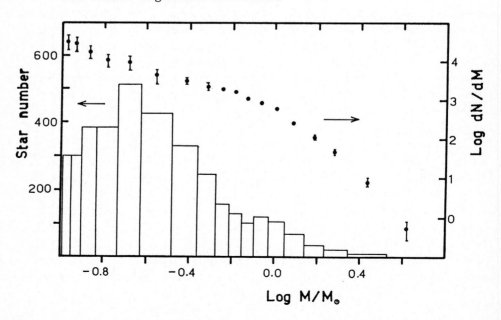

not be neglected for the interpretation of observations. When
predictions on the number densities of stars in a given interval of -for
instance visual- magnitude (that is, the luminosity function LF) are to
be done, one has to remember that the LF is defined as the Mass Function
(MF, number of stars per unit interval of mass) times the derivative of
the mass - visual magnitude relation, that is:

$$dN/dMv = (dN/dM)\times(dM/dlog(L/L\odot))\times(dlog(L/L\odot)/dMv) \qquad (1)$$

Even if we exactly know the bolometric corrections, the LF does not
reflect directly the MF behaviour, if the derivative of the ML relation
is a strong function of the mass, as it is towards the HBMM (DM 1983).
Namely, a decreasing MF does not necessarily means that the MF is
decreasing! DM 1986 have derived the MF:

$$dN/dM = (dN/dMv)\times(dlog(L/L\odot)/dM)\times(dMv/dlog(L/L\odot)) \qquad (2)$$

from the LF of main sequence stars within 20 pc, recently updated by
Wielen et al (1983), together with the ML relation from DM 1985. The MF
is shown in figure 3. The MF at M<1.0M⊙ is also the IMF, as no disk star
of such masses had time to evolve off main sequence. The representation
of the IMF by a power law at its low mass end (0.1<M/M⊙<0.3) gives

$$dN/dM = 316 \ M^{(-2.0\pm0.4)} \qquad (3)$$

Therefore, although the LF decreases at Mv>14, the IMF increases down to
almost the HBMM (0.1M⊙). This is clear from the same figure 3, where the
unit Mv intervals of the LF have been transformed into the corresponding
mass intervals: they are very small towards the HBMM. The figure shows
also that, again, the behaviour of the derivative of the ML relation
produces a dip in the LF at Mv=7, although the MF is monotonic
(Mazzitelli 1972). The conclusion is the following: although we must
expect to see very few red dwarfs (there is no missing mass in red
dwarfs) the IMF for VLMs is a steep increasing function down to the
magnitudes (masses) for which we have informations on the LF, and there
is no observational preclusion that it may continue to raise for masses
below the HBMM.

        In a workshop dedicated to BDs it is legitimate to extrapo-
late the IMF derived for VLMs to masses smaller than the HBMM, to
understand how many BDs we need if we wish them to account for about
0.1M⊙pc^{-3} of dark matter in the disk (Bahcall 1984a and 1984b). As
Salpeter's (1955) index for the IMF (-2.35) is within our range of

uncertainty, we use it, calibrated on the LF at Mv=14:

$$dN/dM = 200 \ (M/M_\odot)^{-2.35} \tag{4}$$

to predict the mass density in BDs, which we want to be:

$$\rho_{dark} = 0.1 \ M_\odot pc^{-3} = \int_{Mmin}^{0.08} M \ (dN/dM) \ dM \tag{5}$$

The integration limit Mmin is thus found to be 0.003M$_\odot$, and the number density results to be 11 BDs pc^{-3} . Thus there should be 6 BDs in a radius of only 0.5pc from the Sun! They should be of very small mass mostly, so that their luminosity is very incospicuous. Nevertheless, their possible dynamical effect on the Oort cloud of comets (e.g Hills 1985) should be investigated in detail.

Our models can be used again to derive the LF in the range of BDs, by equation (1), assuming equation (4) for the IMF. In this case, the bolometric corrections have been wildely extrapolated on the basis of Reid and Gilmore (1984), so that figure 4 is more a suggestion than

Figure 4: Extrapolation of the LF into the BDs region

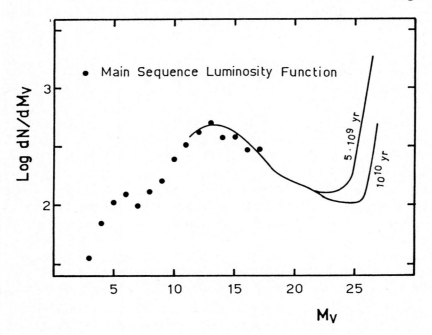

a prediction. After the decline of the LF down to Mv=23, it should rise
again, following the behaviour of the ML relation for thermally cooling
BDs, which is less steep than the ML relation for the last burning
stars.

I wish to stress that the very interesting results presented by Hawkins
(1985, this volume), which seem to indicate a rising LF at Mv=17 are
very difficult to be interpeted in terms of BDs (which should have
really small ages at that magnitude), but mostly may reflect a MF for
stars which has an index even larger than Salpeter's, implying an even
larger number of VLMs and, if extrapolated, of BDs. Viceversa, the very
few VLMs found by Boeshar et al. (1985, this volume) should be interpeted by means of the count predictions of figure 4, and they probably
are still in agreement with a Salpeter's type IMF.

4. THE EVOLUTIONARY STAGE OF VB8B

After the discovery of the infrared companion to VB8 (McCarthy et al
1985), it has become very important to define the age of this star, as
it should be relatively young (5×10^8 to 1×10^9 yr) so that its luminosity ($3 \times 10^{-5} L_\odot$) is compatible with the cooling times of bona fide BDs.
          VB8 is part of a sextuple system including:
i) Wolf 630 (Gliese 644), visual binary with period of 1.7yr, whose B
component is a spectroscopic binary with period of about 1 day (Joy
1947); the C component is VB8.
ii) Wolf 629 (Gliese 643), another visual binary.
This minicluster is the nucleus of the so called Wolf 630 moving group
(Eggen 1977).
          The indicators of an old age are the following:
i) The moving group Wolf 630 is supposed to be as old as M67 (4×10^9 yr -
Eggen 1977);
ii) Wolf 630 itself presents flares, but the H alpha emission core
strength is much smaller than in other similar flare stars, which can be
explained if Wolf 630 is much older (Pettersen et al 1984).
iii) The space velocities components of Wolf 630 with respect to the Sun
are (U,V,W)=(+23,-34,+12)Km s^{-1}(Jahreiss 1985, private communication)
giving a total v=43Km s^{-1} with respect to the Local Standard of Rest.
          Indicators of a young age are:
i) The X-ray luminosity of Wolf 630 is $1. \times 10^{29}$ erg s^{-1}(Johnson 1983).
ii) An H and K index of CaII reversal strength of +7 (on a scale from -
4 to +8) is attributed to Wolf 630 (Wilson and Woolley 1970).
Actually, these latter are not properly indicators of age, but mainly

the fast rotation which young single stars present: in fact there exists
a very good correlation between both the X-ray luminosity and the CaII
emission intensity and the rotation period of stars, almost independent
of the spectral type (Marilli and Catalano 1984). As the B component of
Wolf 630 is a 1 day period binary, and in this situation the rotation of
components is likely to be synchronous with the revolution, it is highly
plausible that both the X-ray and the CaII emissions have to be
attributed to this component, and are due to the fast rotation, and not
to the young age!

      In regard to the indicators of old age, we have to notice
that the reality of moving groups is very questionable (they are
defined on the basis of the V component of the space velocity only)
(see, e.g. McDonald and Hearnshaw 1983), but there is a very good
correlation between the total space velocity with respect to the LSR
and the age of star groups. From Cardini et al (1977) I have taken
figure 5, where it is reported the HR diagram of stars in Gliese's
(1969) catalogue having $40<v<50 \mathrm{Km\ s^{-1}}$. The location of these stars is

    Figure 5: HR diagram of stars in Gliese's catalogue, having
total velocity $40<v<50$ km/s with respect to LSR. Isochrones
by Vandenberg (1985) are superimposed.

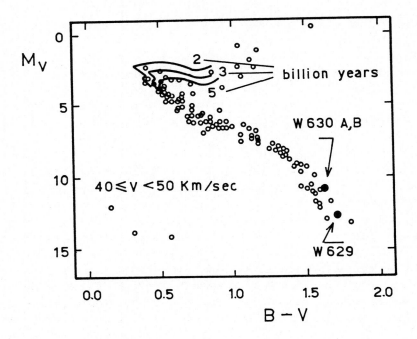

compared with the isochrones of Vandenberg (1985) for solar composition
and ages 2, 3 and 5 billion yr: no stars in the sample is younger than
2 billion years, and probably, from the turn off location, tha average
age of the group is around 5 billion yr. The correlation between total
velocity and age is very good (Cardini et al 1977) also for the other
velocity groups.

I am thus inclined to believe that there are no indications
of youth for VB8, and, on the contrary, consideration of the space
velocity puts a limiting minimum age of 2×10^9 yr for the system. Going
back to figure 1, assigning to VB8B a luminosity of 3×10^{-5} L$_\odot$, it is
clear that its mass must be in the range 0.065 -0.075 M$_\odot$, to be consistent
with model results: in other words, if the age of this system is so
large, VB8B must be partially sustained by nuclear energy sources: it
is a transition mass.

Acknowledgements. Italo Mazzitelli is thanked for a life-long
collaboration and help. I thank also S. Catalano and H. Jahreiss for
very useful informations.

REFERENCES

Alexander,D.R.(1975), Astrophys.J. Suppl. 29,363.
Alexander,D.R.,Johnson,H.R. & Rypma,R.L.(1983) Astrophys. J.
          272 773.
Bahcall,J.N.(1984a), Astrophys.J.,276,169.
Bahcall,J.N.(1984b), Astrophys.J.,287,929.
Boeshar,P.C. & Tyson,J.A.(1985), Astron.J.,90,817.
Boeshar,P.C.,Tyson,J.A. & Seitzer,P.(1985),in this volume.
Cardini,D.,Mazzitelli,I. & Rossi,L.(1977),Astrophys.Space
          Sci.,48,283.
Cox,A.N. & Stewart,J.N.(1970),Astrophys.J.Suppl.,19,243.
D'Antona,F. & Mazzitelli,I.(1982),Astron.Astrophys.,113,303.
D'Antona,F. & Mazzitelli,I.(1983),Astron.Astrophys.,127,149.
D'Antona,F. & Mazzitelli,I.(1985),Astrophys.J.,296,502.
D'Antona,F. & Mazzitelli,I.(1986),Astron.Astrophys.,in press.
Eggen,O.J.(1977), Astrophys.J.,215,812.
Gliese,W.(1969),Catalogue of nearby stars, Astronomischen
          Rechen Instituts Heidelberg,n.22
Graboske,H.C.,DeWitt,H.E.,Grossman,H.S. & Cooper,M.S.(1973)
          Astrophys.J.,181,457.
Grossman,A.S.,Hays,D.,& Graboske,H.C.Jr.(1974), Astron.Astrophys.,
          30,95.

Hawkins,M.R.S.(1985),in this volume.

Hills,J.G.(1985),Astron.J.,90,1876.

Johnson,H.M. (1983), Astrophys.J., 273, 702.

Joy,A.H.(1947),Astrophys.J.,105,96.

Magni,G. & Mazzitelli,I.(19 79),Astron.Astrophys.,72,134.

Marilli,E. & Catalano,S.(1984),Astron.Astrophys.,133,57.

Mazzitelli,I.(1972),Astrophys.Space Sci.,17,378.

Mazzitelli,I. & D'Antona,F.(1986),Astrophys.J.,in press.

McCarthy,D.W.,Probst,R.G.,& Low,F.J. (1985), Astrophys.J., 290,L9.

McDonald,A.R. & Hearnshaw,J.B.(1983),Mon.Not.Roy.Astr.Soc.
        204,841.

Pettersen,B.R.,Evans,D.S. & Coleman,L.A.(1984),Astrophys.J.
        282,214.

Reid,N. & Gilmore,G.(1984),Mon.Not.Roy.Astr.Soc.,206,19.

Salpeter,E.E.(1955),Astrophys.J.,121,161.

Sienkievicz,R.(1982),Acta Astr.,32,275.

Stevenson,D.J.(1978),Proc.A.S.A.,3,227.

Vandenberg,D.A.(1985),Astrophys.J.Suppl.,58,711.

Wielen,R.,Jahreiss,H. & Kruger,R.(1983),in IAU Coll. 76
        'The nearby stars and the stellar luminosity function',
        ed. A.G.Davis Philip and A.R.Upgren (L.Davis Press Inc.,
        Schenectady,New York),p.155.

Wilson,O.C. & Woolley,R.(1970),Mon.Not.Roy.Astr.Soc.,
        148,463.

QUESTIONS

TEPLITZ: In the internal structure calculations for Brown Dwarfs or low
mass stars, are there any periods when they are susceptible of
oscillations (which could provide a characteristic signal?

D'ANTONA: I did not perform any stability analysis.

ZINNECKER: How would the faint end of the theoretical luminosity
function change if you used a Salpeter IMF and a mass luminosity
relation appropriate for less than solar abundance? Does the dip become
less pronounced?

D'ANTONA: You may get a first answer by looking at figure 5 in DM 1982,
which shows the comparison between pop.I and pop.II theoretical
luminosity functions. Actually, the ML relation becomes steeper, and

the HBMM and minimum luminosity increase when decreasing the opacities, so the dip has to become more pronounced. I am presently computing pop.II low mass evolution, and the results will soon be available.

LUNINE: You use opacity tables which assume thermal broadening (in molecular lines of H2O, CO, etc., and weight most highly the lowest opacities (Rosseland mean). In Brown Dwarfs, the pressures are high enough that pressure broadening becomes important in determining the opacity. How much does this affect the interior structures, and hence nuclear burning, of your models?

D'ANTONA: Actually, the models in DM 1985 are not much affected by this problem in the atmosphere, as the densities at the photosphere range from 10^{-5} to 10^{-4}g cm^{-3}. In the subatmosphere, the presence of convection renders less critical a bad determination of opacities. Much more severe problems arise when considering low mass stars or brown dwarfs with reduced metal contents (and opacities): in this case, the photosphere may be even in the critical region where covolume is dominant, and we are close to the pressure ionization boundary: in that case, I am afraid that any opacity presently available may be wrong by orders of magnitude.

PROBST: There is one piece of evidence for a very young age for VB 8 a,b. The primary, Wolf 630, is overluminous in the Mv – V-K diagram by about the L-shift between Grossman's 10^8 and 10^9 yr isochrones. It could be young and still contracting. The same is true for the primary of VB 10.

D'ANTONA: I am not familiar with the Mv – V-K color diagram, but I can refer, as an example, to the Mi – R-I diagram (Eggen 1978, Astrophys.J. 226,405), where VB8 a, for which the pre-main sequence age should be even longer, is on the same color sequence as Proxima Cen, which is probably as old as the Sun. In the Mv – B-V plane, it is well known (see figure 5 and Cardini et al. 1977) that each group of age –subdivided by means of the space velocities– shows a main sequence having an intrinsic spread of about one magnitude, which probably reflects an intrinsic spread in metallicity. I would be very careful in interpreting the position in the color magnitude diagram as an indication of pre-main sequence evolution, as too many other parameters would not fit, in this case (for instance, the low level of flare activity).

EVOLUTION OF SUPER-JUPITERS

W. B. Hubbard
Department of Planetary Sciences, Lunar and Planetary
Laboratory, University of Arizona, Tucson, Arizona 85721

Abstract. In collaboration with J. Lunine and M. Marley we
have carried out a study of the contraction/cooling of degenerate brown
dwarfs in the substellar mass range of 30-100 x 10^{30} g, as well as of
objects in the planetary mass range of 2 x 10^{30} g (Jupiter). The
calculations are based upon an extension of planetary interior physics
to higher densities and pressures, and make use of the thermodynamic
theory for liquid metallic-hydrogen and helium mixtures which we are
currently using to calculate accurate interior models of Jupiter and
Saturn. This theory is based upon three-dimensional Thomas-Fermi-Dirac
(TFD3D) calculations carried out by Hubbard & MacFarlane (1985), which
yield a complete thermodynamic description at finite temperatures. In
this paper we review the physics of the interiors of Jovian-class objects and discuss its extension to substellar objects. Some results are
presented for objects cooling from effective temperatures starting at
2500 K. The interior calculations are coupled to model atmosphere
surface conditions computed by J. Lunine (presented in another paper at
this conference). The latter are continued to higher temperatures and
pressures in the ideal-gas region, and are then matched by entropy to
the strongly-coupled interior region. Except in the planetary-mass
models, this transition region occupies a negligible fraction of the
total mass. Our results are in qualitative agreement with a number of
earlier studies, but are, we believe, more accurate. For the observed
parameters of VB8B, only models with masses in excess of 100 x 10^{30} g
(0.05 solar masses, or 50 Jovian masses) have cooling ages in excess of
10^9 years. Thus for plausible VB8B lifetimes, this object is likely to
be close to the critical hydrogen-ignition mass.

INTERIOR PHYSICS OF JOVIAN-CLASS OBJECTS

For the purposes of this paper, we will consider a Jovian-
class object to be an electron-degenerate self-gravitating object, of
solar or near-solar bulk composition, with interior temperatures below
the range required for significant hydrogen fusion reactions, and with a
mass in the range from ~0.5 x 10^{30} g (Saturn) to ~2 x 10^{30} g (Jupiter).
As we shall discuss, essentially the same set of constituent relations
can be successfully applied to objects with masses up to about two
orders of magnitude higher, or ~100 x 10^{30} g -- such objects can be
called super-Jupiters, or brown dwarfs.

The pressure-density relation for electron-degenerate material at high
pressures takes the limiting form (Salpeter 1961)

$$P_0 = 0.176 \ r_e^{-5} \ [1 - (0.407 \ Z^{2/3} + 0.207) \ r_e], \qquad (1)$$

where all quantities are expressed in atomic units. Here P_0 is the zero-temperature pressure, Z is the atomic number of the degenerate material, and r_e is the density parameter which is related to the volume per electron by

$$v = 4\pi r_e^3/3, \tag{2}$$

where v is the volume per electron. In atomic units, volumes are measured in units of a_0^3, where a_0 is the Bohr radius, lengths are measured in units of a_0, energies are measured in units of e^2/a_0 (e = electron charge), pressures are measured in units of e^2/a_0^4, etc.

To the extent that the correction factor in brackets in eq. (1) can be set equal to unity (i.e., in the limit $r_e \longrightarrow 0$, or electron density becomes very large), the degenerate pressure obeys the standard Chandrasekhar relation for nonrelativistic electrons. In this limit, as is well known, the hydrostatic-equilibrium structure of the object is that of a white dwarf with mass substantially below the Chandrasekhar limit, i.e., a polytrope of index 1.5, with a radius R which varies as $M^{-1/3}$, where M is the mass.

In the mass range of concern here, it is straightforward to verify that typical values of r_e in the object's interior are of order unity, and it is therefore not in general a good approximation to adopt the approximation of a polytrope of index 1.5. Actually, in this mass range, approximation (1) is not highly accurate either. A more general equation of state is really necessary for quantitative work on Jovian-class objects. As is exhibited by eq. (1), it is a general property of thermodynamic relations for degenerate objects that the equations of state approach the ideal Fermi-gas limit as the density and pressure increase. Conversely, progressively more elaborate treatments of the constituent relations are required as the mass decreases from the brown-dwarf range into the Jovian range.

In this study of Jovian and super-Jovian objects, we utilize a generalized thermodynamic description of electron-degenerate hydrogen-helium mixtures which was developed for use in computations of interior models of Jupiter and Saturn (MacFarlane & Hubbard 1983, MacFarlane 1984, Hubbard & MacFarlane 1985). This approach thus guarantees an adequate description of interior physics in the low-mass range of the models considered, while it correctly tends to the limit of a polytrope of index 1.5 in the high-mass range. The calculation of finite-temperature thermodynamics begins with a calculation of the energy and pressure of a lattice of hydrogen and helium nuclei immersed in a degenerate electron gas, using a three-dimensional Thomas-Fermi-Dirac theory (TFD3D). The results are then expressed in terms of effective potentials between the various particles. With explicit potentials available, the finite-temperature thermodynamics of a liquid solar-composition mixture of hydrogen and helium can then be calculated. In particular, the important parameters of the heat capacity, the adiabatic exponent, and the specific entropy can be obtained explicitly, along with the equations of state. In the limit of low temperature and high pressure, the pressure-density relation approaches Salpeter's limiting form, eq. (1).

Deviations from ideality have an important influence on the evolution of our models in at least two respects. First, even at relatively low pressures the effective pressure-density relation behaves qualitatively like (1), in that nonideality reduces the pressure below the value given in the Chandrasekhar limit. The rate of change of pressure with density is correspondingly steepened, with the result that the effective polytropic index is more nearly equal to 1.0, particularly for objects with masses on the order of 10 Jupiter masses and less. As a consequence, there is only a slight dependence of the zero-temperature radius on mass: all of our models approach a similar asymptotic radius which is in the range 55,000 - 80,000 km (Jupiter's radius = 70,000 km and Saturn's radius = 60,000 km). On the other hand, in some cases the radius depends very substantially on the mean interior temperature, because the thermal component can comprise ~10-50% of the total pressure, even under conditions of strong electron degeneracy, because of the relatively low electron densities involved.

The second important influence of nonideality is on the mean heat capacity of the finite-temperature object. The Kelvin time is basically proportional to this quantity, which thus directly affects our estimate of the age of a contracting brown dwarf. Interactions tend to increase the degrees of freedom of protons and helium atoms in a degenerate electron gas, which correspondingly increases the heat capacity. For our models, the temperatures are in the range where the heavy particles (mostly protons and helium nuclei) interact strongly in a liquid phase. At the same time, because of the low electron density, they are substantially shielded by the electrons. Simple analytic models predict that the heat capacity per heavy particle will lie in the range 1.5-3.0 k_B (k_B = Boltzmann's constant), but can seldom do much better.

Phase diagram of hydrogen

Figure 1 shows a schematic phase diagram for hydrogen along with temperature-density relations for Jupiter (J) and two brown-dwarf models (A, B); the latter will be discussed in more detail below. The regime with $r_e < 1$ corresponds to $\log \rho > 0.4$. To the right of this boundary, hydrogen is fully pressure-ionized, and forms metallic hydrogen (denoted as H+) (Hubbard & Smoluchowski 1973; Stevenson & Salpeter 1976). The metallic hydrogen is certainly liquid at temperatures of interest for models discussed here; a solid phase may exist at temperatures and pressures unlikely to be reached by any astrophysical object. At lower pressures (less than about 3 Mbar), solid metallic hydrogen transforms, probably via a first-order phase transition, to a solid phase of molecular hydrogen (denoted as H2). At still lower pressures, the solid molecular hydrogen phase melts to form liquid molecular hydrogen. Again, solid molecular hydrogen is unlikely to exist in any astrophysical object.

The boundary between gaseous and liquid molecular hydrogen is indistinct except at temperatures below the critical temperature at 33 K. Thus the

$$d \ln t / d \ln M = 2.4, \qquad (6)$$

evaluated for $T_e \simeq 1300$ K. The calculated age of a brown dwarf with mass 100×10^{30} g and $T_e \simeq 1300$ K is also dependent on the precise atmospheric boundary condition used. Results using the Planck mean molecular-hydrogen opacity and cloud opacities can be 50-100% larger than results from the Rosseland mean molecular-hydrogen opacity. Nevertheless, it is clear that VB8B can have a cooling age of 10^9 years or greater only if it has a mass greater than or equal to about 0.05 solar masses.

Table 2. Some models of VB8B (Rosseland mean opacities)

M (10^{30} g)	T_e (K)	R (km)	L/L_0	$T_{central}$ (K)	$\rho_{central}$ (g/cc)
30	1317	79460	3.5×10^{-5}	3.0×10^5	62.1

cooling age (time since T_e = 2200 K) = t = 2.7×10^7 yr

M (10^{30} g)	T_e (K)	R (km)	L/L_0	$T_{central}$ (K)	$\rho_{central}$ (g/cc)
50	1317	71370	2.9×10^{-5}	4.9×10^5	150.5

cooling age (time since T_e = 2200 K) = t = 9.1×10^7 yr

M (10^{30} g)	T_e (K)	R (km)	L/L_0	$T_{central}$ (K)	$\rho_{central}$ (g/cc)
100	1317	59620	2.0×10^{-5}	9.3×10^5	568.1

cooling age (time since T_e = 2200 K) = t = 5.2×10^8 yr

CONCLUSIONS

VB8B differs qualitatively from the planet Jupiter. The most plausible model for this object has a mass, central temperature, and central density all of which are about 50 times greater than the corresponding quantities for Jupiter. Thus VB8B is probably more similar to a star than to a planet. Its mass is probably within a factor of two of the minimum mass for hydrogen fusion. We must seek further for Jovian-type planets outside the solar system. Indeed, they will be very difficult to find by photometric methods, since they will be as luminous as VB8B (already the least-luminous known star) only for very brief periods after their formation.

This work was supported by NASA Grant NAGW-192. Mark Marley provided substantial assistance with the calculations.

REFERENCES

Hayashi, C., & Nakano, T. (1963). Evolution of stars of small masses in the pre-main-sequence stages. Prog. Theor. Phys., 30, 460–474.

Hubbard, W. B. (1973). Observational constraint on the structure of hydrogen planets. Astrophys. J. (Lett.), 182, L35–L38.

Hubbard, W. B. (1977). The Jovian surface condition and cooling rate. Icarus, 30, 305–310.

Hubbard, W. B., & MacFarlane, J. J. (1985). Statistical mechanics of light elements at high pressure. VIII. Thomas–Fermi–Dirac theory for binary mixtures of H with He, C, and O. Astrophys. J., 297, 133–144.

Hubbard, W. B., and Smoluchowski, R. (1973). Structure of Jupiter and Saturn. Space Sci. Rev., 14, 599–662.

Lunine, J. I., Hubbard, W. B., & Marley, M. (1986). In preparation.

MacFarlane, J. J. (1984). Statistical mechanics of light elements at high pressure. VI. Liquid-state calculations with Thomas–Fermi–Dirac theory. Astrophys. J., 280, 339–345.

MacFarlane, J. J., & Hubbard, W. B. (1983). Statistical mechanics of light elements at high pressure. V. Three-dimensional Thomas–Fermi–Dirac theory. Astrophys. J., 272, 301–310.

McCarthy, D. W., Jr., Probst, R. G., & Low, F. J. (1985). Astrophys. J. (Lett.), 290, L9–L13.

Salpeter, E. E. (1961). Energy and pressure of a zero-temperature plasma. Astrophys. J., 134, 669–682.

Stevenson, D. J., and Salpeter, E. E. (1976). Interior models of Jupiter. In Jupiter, ed. T. Gehrels, pp. 85–112. Tucson: University of Arizona Press.

DISCUSSION

M. KAFATOS: I would like to ask a simple minded question. Why is 10^9 yr. a reasonable age for VB8B? Could it be several times 10^8 yrs. old?

W. B. HUBBARD: We do not have strong constraints on the age of the system. McCarthy, Probst, & Low (1985) estimate the age to lie in the range 5 x 10^8 to 5 x 10^9 years. There is little evidence that the system was recently formed. In any case it would seem unlikely to me that we should see an object of this type which was formed very recently compared with the age of the local stellar population.

R. BROBST: In the case of VB8B, VB8 seems to be more luminous than the main-sequence luminosity would indicate, and it may be then as young as 10^8 years. All other astrophysical evidence, however, points to $\gtrsim 10^9$ yr. old.

COMPOSITIONAL INDICATORS IN INFRARED SPECTRA OF BROWN DWARFS

Jonathan I. Lunine
Lunar and Planetary Lab., U. of Arizona, Tucson 85721   USA

Abstract. In conjunction with modeling of the evolution of brown dwarfs (see the article in this volume by W. B. Hubbard) we have constructed temperature-pressure profiles in atmospheres of various compositions, for objects ranging from a Jupiter mass up to 0.05 solar masses, and effective temperatures up to 2200K. These profiles are used (1) as boundary conditions to interior models and (2) to construct broad-band emergent infrared spectra of brown dwarfs. We focus here on assessing the compositional information which can be derived from comparison of synthetic spectra with available or anticipated data on the infrared companion to VB8 . We reach the following preliminary conclusions about this object: (1) the shape of the infrared spectrum should be very sensitive to the elemental carbon-to-oxygen ratio in the envelope, because of the oxidation state of carbon; (2) the effective temperature of VB8B, as constrained by the measured flux and modeled radius, must be close to the blackbody temperature of 1360 K calculated by McCarthy, et. al. (1985); (3) the presence of grain opacity in the envelope is hinted at by the available data.

INTRODUCTION

The recent discovery of a low luminosity companion to the late M-dwarf Van Biesbroeck 8 (McCarthy, et. al., 1985) has renewed interest in modeling the physical properties and evolution of objects below the hydrogen burning limit of roughly 0.08 solar masses ($M_\odot$) (Grossman, et. al., 1974; D'Antona & Mazzitelli, 1985), down to Jupiter-mass bodies. To completely characterize so-called brown dwarfs, the observed spectral distribution of flux and inferred or "guessed" age must be matched by a model in which mass, radius, effective temperature and composition are determined in a self-consistent manner. This requires that physically reasonable models of both the interior and outer envelope (hereafter "atmosphere") be constructed and coupled together. In collaboration with W.B. Hubbard and M. Marley we have undertaken such an effort (Lunine, et. al., 1986) to characterize the nature of brown dwarfs over the mass range cited above, and in particular to derive as much information as possible from the existing observations of the object VB8B. The construction of interior models is described by Hubbard elsewhere in this volume. Here we consider the calculation of the atmospheric temperature structure, composition of the gaseous and condensed molecular species present for a given elemental composition, and the resulting emergent infrared spectra.

CONSTRUCTION OF MODEL ATMOSPHERES

We define the atmosphere as the outermost portion of the brown dwarf structure in which unity optical depth is first reached and surpassed at optical to infrared wavelengths. It is this region which provides the outer boundary condition for the internal temperature structure. It is also from this region that the infrared and visible radiation are emitted, and hence compositional information can be derived, as has been done for stars and the outer planets of our solar system. In constructing model atmospheres, we make the following assumptions: (1) The object is predominantly hydrogen and helium in solar proportions. The abundances of heavier elements such as oxygen, carbon and nitrogen are allowed to vary but are of order the solar values. (2) A primarily molecular hydrogen ideal gas is assumed for the atmosphere, which is valid up to about 2500 K temperature. (3) The atmosphere is plane parallel and isotropic, and hydrostatic equilibrium applies. (4) The atmosphere is in local thermodynamic equilibrium down to the level at which the temperature gradient becomes steep enough that the gas is unstable against convection. (5) The grey atmosphere solution to the equation of radiative transfer in the Eddington approximation (Chandrasekhar, 1960, p. 294), along with "suitably" chosen frequency-averaged opacities, yields adequate bounds on the temperature structure. (6) The interior is fully convective and isentropic. Justifications and implications of the assumptions are discussed in Lunine, et. al. (1986).

Given the above assumptions along with an assumed mass, radius, composition and effective temperature, the temperature structure (run of temperature with pressure) of the atmosphere is evaluated using both Rosseland and Planck mean opacities (Mihalas, 1978, p.56). The model atmosphere determines the conditions at the top of the convection zone which is used as a boundary condition for the interior model. Consistency between the resulting structure and the initially assumed parameters is then imposed by adjusting the input parameters and rerunning the model. The flux as a function of frequency is calculated using the temperature structures derived in the Rosseland and Planck mean limits, along with frequency-dependent opacities used as input to the opacity-mean calculations. The effective temperature T_e is calculated from the emergent spectrum, and compared to the input value. In general, the output T_e is less than the input in the Rosseland mean case, and greater in the Planck case. The "correct" temperature structure which would be derived by a frequency-dependent solution of the radiative transfer equation is bounded by the two grey cases.

The primary opacity sources considered are pressure-induced molecular hydrogen (augmented by helium), H^-, H_2^-, CO, CH_4, H_2O, and grains. The pressure-induced opacity is from Linsky (1969) and Patch (1971); the opacity from ions is taken from Gingerich (1964), Linsky (1969), and Vardya (1965). The molecular band absorptions of CO, CH_4 and H_2O are calculated from an Elsasser model described in Edwards, et. al. (1967) and tied to laboratory data of Ludwig (1971). The laboratory data cover a pressure-temperature range comparable to that of the VB8B model (see below). Based upon our results the dominant nitrogen species in the atmosphere of brown dwarfs with effective temperatures in excess of several hundred degrees is N_2 rather than NH_3 as in the Jovian planets; hence nitrogen opacity is unimportant. Grain opacity, due to condensed silicates and iron (see the cloud models below), is incorporated using the absorption coefficients of Pollack, et. al. (1973).

The construction of frequency-dependent opacity models based upon the above sources, as well as the results of the interior models, enables one to draw the important conclusion that in modeling objects in the mass regime of 0.01-0.05$M_\odot$ and effective temperature of 1000-2000K, the use of frequency-averaged opacity tables from stellar work such as that of Alexander, et. al. (1983) is not valid. There are three major problems with extrapolating from such tables, all tied to the low temperatures and high surface gravities (10^4-10^5 cm s^{-2}): (1) line widths and profiles in the bands of the molecular species are due primarily to pressure-broadening, rather than thermal broadening as assumed in the tables; (2) pressure-induced hydrogen absorption is much more important in brown dwarfs than in the hotter M-dwarfs, and (3) realistic silicate cloud models are required to describe correctly the effect of condensates on brown dwarf atmospheres, and these are not included in tables.

Construction of the temperature profiles and comparison with condensation curves of various molecules indicates that in a solar composition atmosphere, with effective temperature of 1500 K and ϵ surface gravity of 10^5 cm s^{-2} (mass 0.05 $M_\odot$, radius 0.1 $R_\odot$), iron and various silicates will condense out near the gas unity optical depth level. These condensates have a dominant effect on the thermal structure, comparable to the role of ammonia and water clouds in the atmospheres of Jupiter and Saturn. Two extreme cloud models are considered: (1) Condensation clouds, in which the grain mass fraction follows the saturation vapor pressure law (Barshay & Lewis, 1976); (2) dusty clouds, in which condensate is transported upwards by vigorous convection from the cloud forming region, and falls off as gas pressure squared (Prinn & Olaguer, 1981; Lunine, et. al., 1986). The two represent "thin" and "thick" cloud extremes.

EFFECT OF CARBON TO OXYGEN RATIO ON SPECTRA

Figure 1 shows two representative temperature profiles for a brown dwarf with effective temperature of 1500K, mass 0.05 $M_\odot$, and radius 0.1 $R_\odot$. We focus on this sort of object because, as discussed in the article by Hubbard in this volume, it retains a sufficiently high luminosity over 10^9 years to be observable with existing or anticipated techniques. As discussed in the following section, VB8B is likely to be in this mass and effective temperature range. The two profiles A and B correspond to models with and without thick clouds, respectively.

Superimposed are dashed lines showing the transitions between oxidized
and reduced species of carbon and nitrogen in the brown dwarf envelopes,
for solar composition from Barshay & Lewis (1976). Also shown for
comparison are an adiabat for the deep interior of Jupiter and a
temperature profile for a model solar nebula, from Barshay & Lewis
(1976).

The fundamental chemical difference between high mass brown dwarfs and
Jovian planets exhibited by the figure is that CO dominates in the
former. As a consequence, the abundance of gaseous water is a strong
function of the C to O ratio in the atmosphere, for values near solar.
The effect of water on the emergent spectrum is illustrated in figure 2,
which shows an atmosphere with a C to O ratio less than and greater than
unity, corresponding respectively to the presence and absence of H_2O.
Broadband infrared spectra, particularly if wavelength regions outside
the telluric water bands are observed, are potentially very sensitive to
an important elemental ratio, that of carbon to oxygen, in the
atmospheres of brown dwarfs such as VB8B. In contrast, determination of
the C to O ratio in the Jovian and Saturnian atmospheres, where these
two elements are not coupled in molecular species, has proved difficult.

Figure 1. Temperature profile in brown dwarf atmospheres,
the deep Jovian atmosphere and solar nebula model. A and B
refer to brown dwarf models with and without clouds. Dashed
lines indicate boundaries between oxidized and reduced
carbon and nitrogen species. All lines except A and B from
Barshay and Lewis (1976).

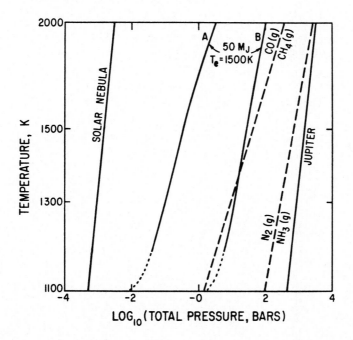

EFFECTIVE TEMPERATURE OF VB8B

The coupled atmospheric and interior models have been used to further our understanding of the brown dwarf candidate VB8B. The interior models tightly constrain the radius to be between 0.09 and 0.12 $R_{\odot}$, which permits us to fit directly the magnitudes of the H and K band fluxes measured by McCarthy, et. al. (1985). A general conclusion that can be made is the following: because the radius derived by McCarthy et. al. (1985) from their data, assuming a blackbody, is close to the physical radius derived in our interior models, the spectrum of VB8B in the one to three micron wavelength range must either be (1) nearly grey and have an effective temperature close to the 1360 K derived by McCarthy, et. al., or (2) coincidentally mimic a greybody in the H and K bands. The latter possibility is unlikely considering the broadness of the bands and likely absorbers. We find that we can fit the H and K band fluxes with an effective temperature in the range 1200 to 1400 K. This fit requires, however, that a grey absorber be present in the atmosphere. The data cannot be fitted with an atmosphere containing molecular absorbers only, because these fall off sharply in opacity from one to three microns, making the spectrum rather non-grey. The most reasonable candidate is silicate grain opacity, which is fairly uniform over the relevant wavelength range (Pollack, et. al. 1973), and would be expected to condense in a solar composition atmosphere under the model conditions of temperature and pressure. We therefore propose, based upon preliminary fitting of H and K fluxes to model emergent spectra as well as modeling of the cooling history (see the Hubbard article), that VB8B is an object of roughly 0.05 $M_{\odot}$ with an effective temperature in the range 1200 to 1400 K, and an atmosphere containing at least one to several optical thicknesses of silicate (and possibly iron) grains.

Figure 2. Emergent spectra of brown dwarf with carbon to oxygen ratio less than (left panel) and greater than (right panel) unity.

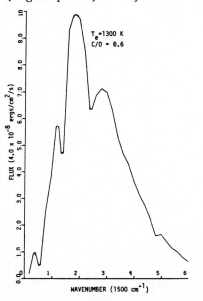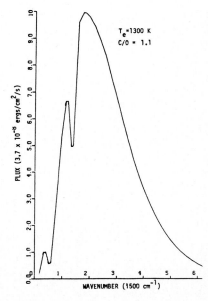

REFERENCES

Alexander, D.R., Johnson, H.R. & Rypma, R.L. (1983). Effect of molecules and grains on Rosseland mean opacities. Astrophys. J., 272, 773-80.

Barshay, S.S. & Lewis, J.S. (1976). Chemistry of primitive solar material. Ann. Reviews Astron. Astrophys., 14, 81-94.

Chandrasekhar, S. (1960). Radiative Transfer. New York: Dover.

D'Antona, F. & Mazzitelli, I. (1985). Evolution of very low mass stars and brown dwarfs. I. The minimum main sequence mass and luminosity. Astrophys. J., 296, 502-13.

Edwards, D.K., Glassen, L.K., Hauser, W.C. & Tuchscher, J.S. (1967). Radiation heat transfer in nonisothermal nongray gases. J. Heat Transfer, 89, 219-29.

Gingerich, O. (1964). Review of opacity calculations. In Proc. First Harvard-Smithsonian Conf. on Stellar Atmospheres. pp. 17-41. Smithsonian Astrophys. Obs. Spec. Rep. 167.

Grossman, A.S., Hayes, D. & Graboske, H.C., Jr. (1974). The theoretical low mass main sequence. Astron. & Astrophys., 30, 95-103.

Linsky, J.S. (1969). On the pressure-induced opacity of molecular hydrogen in late-type stars. Astrophys. J., 156, 989-1005.

Ludwig, C.B. (1971). Measurements of the curves-of-growth of hot water vapor. Applied Optics, 10, 1057-73.

Lunine, J.I., Hubbard, W.B. and Marley, M.S. (1986). The evolution and physical nature of brown dwarfs. Astrophys. J., submitted.

McCarthy, D.W., Jr., Probst, R.G. & Low, F.J. (1985). Infrared detection of a close cool companion to Van Biesbroeck 8. Astrophys. J., 290, L9-13.

Mihalas, D. (1978). Stellar Atmospheres. San Francisco: W.H. Freeman.

Patch, R.W. (1971). Absorption coefficients for hydrogen. II. Calculated pressure-induced H_2-H_2 vibrational absorption in the fundamental region. J. Quant. Spectrosc. Radiat. Transfer, 11, 1331-53.

Pollack, J.B., Toon, O.B. & Khare, B.N. (1973). Optical properties of some terrestrial rocks and glasses. Icarus, 19, 372-89.

Prinn, R.G. & Olaguer, E.P. (1981). Nitrogen on Jupiter: a deep atmospheric source. J. Geophys. Res., 86, 9895-9.

Vardya, M.S. (1964). Thermodynamics of a solar composition gaseous mixture. Astrophys. J., 129, 205-13.

DISCUSSION

Question: (H. Shipman) Since the quoted masses (and gravities) for VB8B vary by a factor of 3, have you explored different gravities?

Answer: (J. Lunine) The main effect of varying the surface gravity over the range suggested is to move the base of the cloud deck, and hence opacity, relative to that of the gas absorbers. We plan to construct more sophisticated cloud models to examine this and other condensate processes.

Question (B. Campbell) Observational difficulties aside, would there be significant advantage to looking for these objects at 4.7 microns?

Answer (J. Lunine) The modest molecular hydrogen opacity dropoff in the 5 micron region gains one only a factor of two in flux over the grey case for high luminosity brown dwarf models. This is in contrast to Jupiter, where a number of important absorbers (such as ammonia) in addition to molecular hydrogen have steep opacity dropoffs which contribute to the great increase in flux.

Question (J. Bahcall) Are there any characteristic spectroscopic features that might be detectable with future space instruments?

Answer (J. Lunine) Bands of water, carbon monoxide and/or methane are key features which spaceborne infrared instruments should be able to detect at low spectroscopic resolution (5 %). Synthetic spectra at higher resolution should be constructed to assess what other, minor constituents would be detectable.

THE EVOLUTION OF VERY LOW-MASS STARS

L. A. Nelson, S. A. Rappaport, and P. C. Joss

Department of Physics, Center for Space Research,
and Center for Theoretical Physics
Massachusetts Institute of Technology

We present the results of numerical evolutionary calculations for low-mass stars in the mass range of $0.01 - 0.10 \, M_\odot$. We have followed the evolution of these stars from the early stages of contraction, through deuterium burning, to the very late stages of degenerate cooling at ages comparable to that of the Galaxy. By varying the surface boundary conditions, we systematically explored the sensitivity of these evolutionary sequences to the major uncertainties in the input physics. We find that, at a given age, the effective temperatures and bolometric luminosities of such stars are quite well determined despite these uncertainties. Our calculations are particularly appropriate to the interpretation of the observations of substellar objects such as Van Biesbroeck 8B. In particular, we find that if VB 8B has an age in excess of $\sim 10^9$ yr, it has a mass in the range of $0.04 - 0.08 \, M_\odot$.

I. INTRODUCTION

There has recently been considerable renewed interest in the subject of brown dwarfs because of (i) the possibility that these objects comprise a significant fraction of the local missing mass in the Galactic disk (see, e.g., Bahcall 1984), and (ii) the recent discovery of the first *bona fide* brown dwarf, VB 8B (McCarthy, Probst, and Low 1985). We take, as a working definition of a brown dwarf, a star that has insufficient mass to achieve thermal equilibrium through hydrogen burning (i.e., mass $M \lesssim 0.08 \, M_\odot$; see, e.g., Kumar 1963; Grossman, Hays and Graboske 1974, hereafter GHG; Rappaport and Joss 1984). The lower mass limit for such objects is not precisely defined, but we consider here only stars with masses sufficiently large to be supported primarily by thermal or electron degeneracy pressure, as opposed to atomic and molecular forces (i.e. mass $M \gtrsim 0.01 \, M_\odot$). Theoretical cooling curves for brown dwarfs are important in the interpretation of observed substellar objects, as well as for a guide in planning future searches for such objects.

The first estimates of the cooling of brown dwarfs were made by Tarter (1975), based on fits to model calculations (e.g., those of GHG) that were then extrapolated to lower effective temperatures ($T_e < 1200$ K) and older ages than were covered by the models. The first calculation of the cooling of brown dwarfs to very low temperatures and old ages was carried out by Stevenson (1978), based on an analytic model wherein the star was taken to have a constant radius and to be highly degenerate. More recently, Nelson, Rappaport, and Joss (1985a; hereafter NRJ) carried out the first self-consistent numerical evolutionary calculations of brown dwarfs including an early phase of stellar contraction, a deuterium-burning main-sequence phase, and finally cooling to a completely degenerate state at ages greater than that of the Galaxy.

In the present work, we have somewhat refined these calculations and, perhaps more importantly, we have systematically explored the sensitivity of the models to the uncertainties in the input physics. We also present more details of the evolution of the stellar interiors than were given in NRJ. In Sect. II we describe the model and the input physics. In Sect. III we present the results of the evolutionary calculations. Our basic results and conclusions are summarized in Sect. IV.

II. THE MODEL

We utilize a simplified stellar evolution code with the following basic properties: (1) The structure of the stellar interior is taken to be that of an $n = 3/2$ polytrope. (2) The atmosphere is represented by the simple pressure boundary condition, $P\kappa = g$, where P, κ, and g are the pressure, radiative opacity, and gravitational acceleration at the photosphere, respectively. (3) The specific entropy in the interior is matched directly to that at the photosphere, thus bypassing the complicated envelope wherein partial ionzation and dissociation zones are found. The basic evolution code has previously been utilized extensively for binary stellar evolution calculations (see, e.g., Rappaport, Joss, and Webbink 1982, hereafter RJW; Rappaport and Joss 1984) and for preliminary calculations of brown-dwarf evolution (NRJ). A similar approach has previously been used by Hubbard (1973) to evolve models of the Jovian planets. The assumption of an $n = 3/2$ polytropic structure is an excellent approximation for times during the evolution when the star is nearly completely convective and for times when the star has significant electron degeneracy. We show later in this paper that the $n = 3/2$ polytropic structure is essentially always valid, since the convective and degenerate stages substantially overlap. We effectively explore the sensitivity of our results to the use of a simple pressure boundary condition at the same time that we test a wide range of possible atmospheric opacities. The sensitivity of our results to the assumption of matching entropies is also extensively studied in this work.

In addition to the pressure due to an arbitrarily degenerate electron gas and an ideal gas of ions, we also include a small approximate correction to take into account Coulomb interactions among the constituent particles. Since the fractional contribution of the Coulomb interactions to the pressure is not constant throughout an $n = 3/2$ polytrope, we have used an average value of the Coulomb contribution based on the mean density of the star. For small values of the plasma parameter Γ (ratio of Coulomb to kinetic energy densities), we utilized the Debye-Hückel formulation to calculate the correction to the perfect-gas equation of state, while for larger values of Γ we used the fomulation of Slattery, Doolen, and DeWitt (1980). Other contributions to the equation of state from, for example, electron exchange and quantum ion effects were not included, due to their relative unimportance in the range of densities and temperatures of interest.

In our "standard model" the stellar evolution is started with a radius $R \simeq R_\odot$, which is very large compared to the star's ultimate radius at old ages. We assume a solar composition ($X = 0.7, Z = 0.02$) plus a primordial abundance of deuterium of 5×10^{-5} by mass (Gautier and Owen 1983). Nuclear burning, including weak and strong screening corrections, are incorporated as described by Rappaport, Verbunt, and Joss (1983) and Rappaport and Joss (1984). In the present calculations, the He^3-burning portion of the p-p chain was excluded in calculating the nuclear energy generation rate, because of the low internal temperatures of these low-mass stars. The specific entropy in the interior, including a correction for the contribution from Coulomb interactions, was matched directly to the specific entropy at the photosphere. Atmospheric radiative opacities were calculated from a fitting formula to the Alexander opacities (Alexander 1975; RJW). For low temperatures (i.e., $T_e \lesssim 2500$ K), the atmospheric radiative opacities become difficult to estimate accurately because of the contribution from molecules and the possible formation of grains. We therefore chose to simply parameterize the opacity coefficients at low temperatures by adopting a constant limiting value for κ (denoted by κ_{min}), which we vary from one model to another. In our standard models, we take the limiting value of the opacity coefficient to be $\log \kappa_{min} = -1.5$.

Much of the uncertainty in the evolution due to the possible existence of nonadiabatic zones in the stellar envelope can be incorporated into a single parameter, namely, the difference in specific entropy, Δs, between the stellar interior (s_i) and that at the photosphere (s_s). Such a fomulation can also take into account many of the other uncertainties in the microphysics of the stellar envelope and stellar interior. We have chosen three algorithms for mismatching the entropy in addition to our standard model with $s_s - s_i \equiv \Delta s = 0$: (i) $\Delta s = +1$; (ii) $s_s = 5/4\,s_i$; and (iii) $\Delta s = -1$ (where we have expressed specific entropy in the natural units of Boltzmann's constant divided by the atomic mass unit). We consider this range of entropy mismatches to be fairly encompassing based on (a) a few of our own Henyey models of stars at low effective temperatures ($2500 \gtrsim T_e \gtrsim 1600$ K; Nelson 1985), and (b) the fact that s varies only from 16 to 6 throughout the course of the evolution in any of our standard models (covering a range of nearly a factor of 10^4 in central density).

To simulate the uncertainties in the atmospheric radiative opacities due to the formation of grains and molecules at low temperatures, we have taken the opacity coefficient to have a constant value κ_{min}, once the coefficient given by the RJW fitting formula (model A) to the Alexander opacities (1975) decreases below κ_{min}. We computed evolutionary sequences for $\log(\kappa_{min}/cm^2\,g^{-1}) = -1.0, -1.5, -2.5,$ and -3.5, as well as a sequence which used the fitting formula A, of RJW, extrapolated to arbitrarily low effective temperatures.

III. RESULTS

The evolution of the stellar radius for our standard models is shown in Figure

1. We carried out evolutionary calculations for masses of 0.01, 0.02, 0.04, 0.06, 0.08, and 0.10 $M_\odot$. For young ages, the stellar radius contracts with elapsed time, t, approximately as $t^{-1/3}$, as can be demonstrated analytically. (We note that for very young ages, i.e., $t < 10^6$ yr, the evolution is highly uncertain due to the possible effects of rapid stellar rotation and mass loss or accretion.) For ages near 10^6 years, there is an interval on each evolutionary curve (for $M \gtrsim 0.015\,M_\odot$) where the radius remains nearly constant; this corresponds to the deuterium-burning phase (see also GHG). After deuterium exhaustion the star contracts, and for a period of time the internal temperatures increase (in accordance with the virial theorem). For stars with masses in excess of $\sim 0.084\,M_\odot$ the internal temperatures reach sufficiently high values to establish thermal equilibrium through thermonuclear burning via the p-p chain. For lower-mass stars, thermal equilibrium is never reached, and the internal temperatures ultimately decline as the electron degeneracy increases. These low-mass stars are destined to cool forever toward a completely degenerate configuration. Note that for old ages ($t > 10^9$ yr) the mass-radius relation is in approximate agreement with that found by Zapolsky and Salpeter (1969) for low-mass, fully degenerate stars (i.e., $R \propto M^{-1/6}$).

The evolution of the effective temperature for our standard models is presented in Figure 2. The temperatures during the contraction phase remain relatively constant (see also, for example, Hayashi and Nakano 1963) and then decline after much of the available gravitational energy has been exhausted. Since all the stars with masses in the range of $0.01 - 0.08\,M_\odot$ eventually attain a similar radius (to within $\pm\,20\%$), the higher mass stars, which initially have a larger store of potential energy ($\propto M^2$), require significantly longer times to cool to the same effective temperature. For old ages ($\gtrsim 5 \times 10^8$ yr) the effective temperatures are given approximately by

$$T_e \simeq 2050 \left(\frac{M}{0.1\,M_\odot}\right)^{0.63} \left(\frac{t}{10^9\,\mathrm{yr}}\right)^{-0.29} \mathrm{K} \;. \tag{1}$$

The results shown in Figures 1 and 2 can be combined to yield the evolution of the stellar luminosity, L, in our standard models (see Figure 3). For old ages, L is given approximately by

$$\left(\frac{L}{L_\odot}\right) \simeq 7.8 \times 10^{-5} \left(\frac{M}{0.1\,M_\odot}\right)^{2.2} \left(\frac{t}{10^9\,\mathrm{yr}}\right)^{-1.2} . \tag{2}$$

Equations (1) and (2) are in good agreement with the earlier work of Stevenson (1978) as well as more recent calculations presented at this conference (see, e.g., Stevenson 1985).

As discussed above, we also carried out a series of evolutionary runs designed to test the sensitivity of our models to the assumption of matching entropies. The

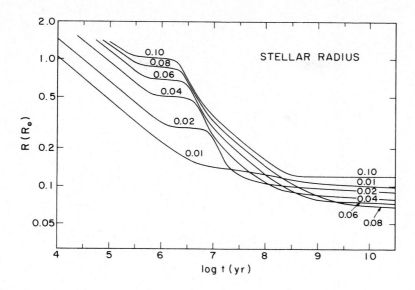

Figure 1: Evolution of stellar radius, R, as a function of age, t, for the "standard" models of very low-mass stars with masses in the range of 0.01–0.10 $M_\odot$. Each track is labelled with the corresponding stellar mass in units of solar masses. (All figures in this article are taken from Nelson, Rappaport, and Joss 1985b.)

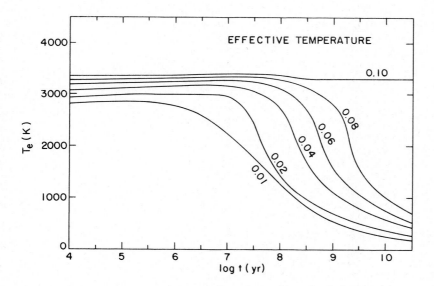

Figure 2: Evolution of the effective temperature, T_e, as a function of age t, for the "standard" models of very low-mass stars with masses in the range of 0.01–0.10 $M_\odot$.

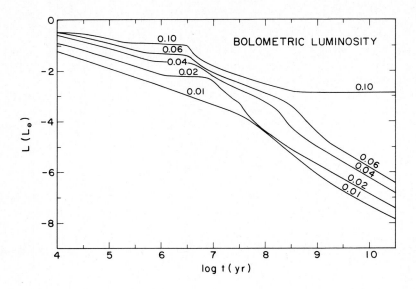

Figure 3: Evolution of the bolometric luminosity, L, as a function of age, t, for the "standard" models of very low-mass stars with masses in the range of 0.01–0.10 $M_\odot$.

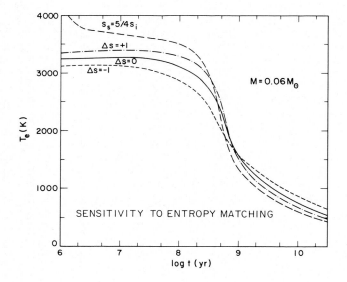

Figure 4: Sensitivity of the cooling of a 0.06 $M_\odot$ stellar model to various mismatches in the specific entropy, s, between the stellar interior and stellar surface. Each track is labelled according to the algorithm by which specific entropy is mismatched; see text.

evolution of T_e for a 0.06 $M_\odot$ star with various degrees of entropy mismatches (see above for details) is shown in Figure 4. Note that for the models with $\Delta s > 0$ the effective temperatures are higher early in the evolution than for the model with $\Delta s < 0$. At older ages the reverse of this condition obtains. This reversal of temperatures is a simple consequence of the virial theorem and the fact that the total energy to be extracted from the stellar contraction is nearly fixed. For $t < 2 \times 10^9$ yr, the uncertainty in T_e at any given time (defined as the half width between extreme model values) is less than 10%. For $t > 2 \times 10^9$ yr the uncertainty in temperature remains fairly constant at $\sim$ 150 K. The corresponding uncertainty in the evolution time required for a star of mass 0.06 $M_\odot$ to cool to a given temperature is less than a factor of 2 for all tested variations of entropy mismatch. A similar series of tests involving entropy mismatches of a star with $M = 0.02\,M_\odot$ is shown in Figure 5. In this case the uncertainty in the temperature is always $< 15\%$, and for $t > 10^8$ yr the uncertainty is only $\sim$ 80 K. The uncertainty in cooling times for this case is less than a factor of 1.6.

Evolutionary tracks for our 0.02 and 0.06 $M_\odot$ models in the central-density/central-temperature ($\rho_c - T_c$) plane are shown in Figure 6. For each stellar mass, the evolutionary track is marked with different values of the entropy mismatch. The track is labelled where the evolution of a particular model sequence terminates ($t = 3 \times 10^{10}$ yr). Note that once M and ρ_c are specified in our models, T_c is uniquely defined; hence, for a given mass, there is only a single track in the $\rho_c - T_c$ plane. However, the surface boundary conditions determine the time required to reach any given point along an evolutionary track. Also indicated in the figure are several contours of constant electron degeneracy (as measured by the value of $F_{1/2}$, the Fermi-Dirac integral of index $1/2$) as well as contours of constant values of the mean plasma parameter Γ. We find that the internal temperatures at old ages vary significantly (by a factor of $\sim$ 4) over the range of assumed entropy mismatches. On the other hand, the largest value of Γ encountered at the stellar center is $\sim$ 15. Thus, although the values of Γ will increase outward from the stellar center, the models we have considered will not undergo crystallization over any significant fraction of their volume for ages up to the age of the Galaxy. Consequently, Debye cooling should be unimportant in these low-mass stars (cf. Tarter 1975).

The sensitivity of our models to uncertainties in the atmospheric radiative opacities is shown in Figures 7, 8, and 9. The evolution of T_e for a star with $M = 0.06\,M_\odot$ and for a wide range of opacity laws is shown in Figure 7. As discussed above, the opacity coefficients were taken from the fitting formula of RJW (Model A), but with various lower limits, $\kappa_{\min}$, imposed. The effects on the evolution of different limiting values for the radiative opacity coefficient were studied over a range of nearly four orders of magnitude in $\kappa_{\min}$. We find that for a star of mass 0.06 $M_\odot$, this wide range of opacity coefficients yields uncertainties in the effective temperatures of only $\lesssim 12\%$ for all ages. The corresponding uncertainty in the cooling times are a factor of $\sim$ 1.5.

The sensitivity to variations in the minimum opacity coefficient, $\kappa_{\min}$, for a 0.02 $M_\odot$ stellar model is shown in Figure 8. The uncertainties in effective temperature at a given age are, as in the case of the 0.06 $M_\odot$ model, less than $\sim$ 12%. The corresponding uncertainties in the cooling times are also quite

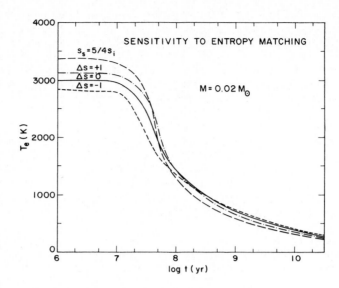

Figure 5: Sensitivity of the cooling of a $0.02\,M_\odot$ stellar model to various mismatches in the specific entropy, s, between the stellar interior and stellar surface. Each track is labelled according to the algorithm by which specific entropy is mismatched; see text.

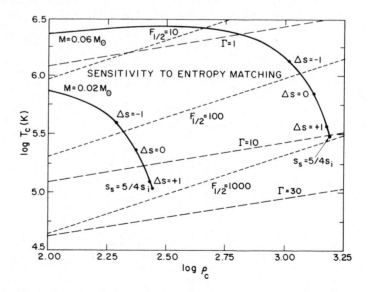

Figure 6: Sensitivity of the central density, ρ_c, and central temperature, T_c, of $0.02\,M_\odot$ and $0.06\,M_\odot$ stellar models to various types of mismatches in specific entropy (see text for details). Each heavy dot on the two curves denotes the internal properties of a particular model at an age of 3×10^{10} yr; $F_{1/2}$ and Γ are the Fermi-Dirac integral of index $1/2$ and the plasma parameter, respectively.

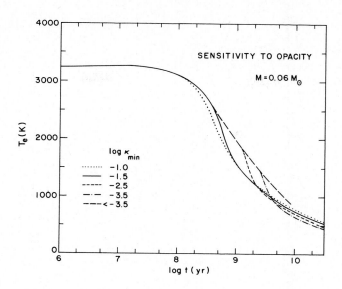

Figure 7: Sensitivity of the cooling of a $0.06\,M_\odot$ stellar model to various assumed minimum values for the atmospheric radiative opacity coefficient, $\kappa_{\min}$.

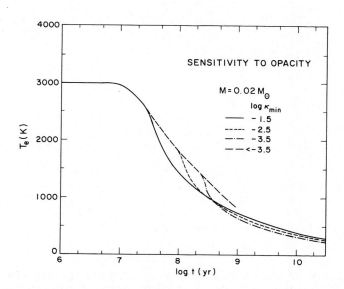

Figure 8: Sensitivity of the cooling of a $0.02\,M_\odot$ stellar model to various assumed minimum values for the atmospheric radiative opacity coefficient, $\kappa_{\min}$.

small, similar to those for the $0.06\,M_\odot$ model.

Evolutionary tracks in the $\rho_c - T_c$ plane for the $0.02\,M_\odot$ and $0.06\,M_\odot$ models, with various values of κ_{min}, are shown in Figure 9. The points on the evolutionary tracks marked with different values of κ_{min} indicate the final evolutionary state at an age of 3×10^{10} yr. The largest value of Γ at the center of any of these models is ~ 30. However, the values of κ_{min} corresponding to the largest values of Γ are quite unrealistic. More plausible values of κ_{min} ($\gtrsim 10^{-2} \mathrm{cm^2\,g^{-1}}$) yield maximum central values of $\Gamma \simeq 10$. Thus, as in our study of possible entropy mismatches (described above), we find that crystallization and Debye cooling will be unimportant for any plausible low-temperature opacity law.

Finally, as a self-consistency check, we monitored the interior of our models for convective instability. Conductive opacities for the stellar interior were taken from fitting formulae derived by Iben (1975) based on the results of the conductive opacity calculations of Hubbard and Lampe (1969). The results of these calculations are shown in Figure 10, which displays the mass fraction of the star wherein a conductive core would be likely to form within the convective envelope. (We have not yet, however, calculated fully self-consistent stellar models containing conductive cores, and the mass fractions for these cores, as shown in the figure, may well be overestimates.) Note that the models we consider would be completely convective for $t \lesssim 2 \times 10^9$ yr in the higher-mass stars and $t \lesssim 5 \times 10^9$ yr in the lower-mass stars. For an age equal to that of the solar system, stars with $M \lesssim 0.03\,M_\odot$ are almost completely convective, whereas for higher mass stars a conductive core might occupy roughly half the mass. However, it is important to note that by the time these stars develop significant conductive cores, they are quite degenerate. Hence, the assumption of an $n = 3/2$ polytropic structure is still extremely good even though the temperature gradient may be small through the central portion of the star. Thus, the principal effect of the onset of a conductive core should be only a slight redistribution of the luminosity with respect to age.

IV. CONCLUSIONS

From our evolutionary calculations of low-mass stars in the range of 0.01 to $0.08\,M_\odot$ we were able to reach a number of secure conclusions:

1. The evolution of the effective temperature and bolometric luminosity are fairly well determined in spite of the present uncertainties in the input physics. In particular, we have shown that the evolution is remarkably insensitive to the choice of the atmospheric opacity law at low temperatures and to the amount of mismatch in specific entropy across the stellar envelope.

2. The uncertainties in the effective temperatures for stellar ages up to that of the Galaxy are typically only $\pm 10\%$. However, as pointed out by a number of workers at this conference (e.g., Bahcall 1985; Lunine 1985), there are much greater theoretical uncertainties in the spectral features of such low-temperature stars even if the effective temperature were known exactly.

3. For evolutionary ages in excess of 5×10^8 yr, the effective temperature and

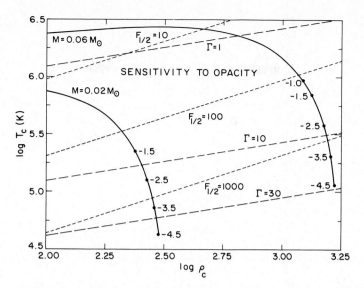

Figure 9: Sensitivity of the central density, ρ_c, and central temperature, T_c, of $0.02\,M_\odot$ and $0.06\,M_\odot$ stellar models to various assumed minimum values for the atmospheric radiative opacity coefficient, $\kappa_{\min}$. Each heavy dot on the two curves denotes the internal properties of a particular model [denoted by the value of $\log(\kappa_{\min}/\mathrm{cm}^2\,\mathrm{g}^{-1})$] at an age of 3×10^{10} yr; $F_{1/2}$ and Γ are the Fermi-Dirac integral of index $1/2$ and the plasma parameter, respectively.

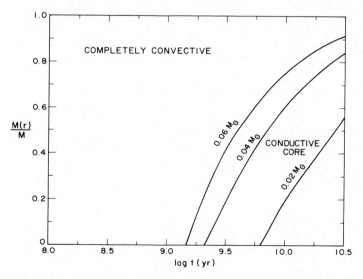

Figure 10: Preliminary estimates of the fraction of the stellar mass, (M_r/M), for which electron conduction would dominate the energy transport as a function of age. Results are shown for 0.02, 0.04, and 0.06 $M_\odot$ models. The remainder of the stellar interior is unstable against convection.

luminosity can be reasonably well represented by power-law fitting formulae with $T_e \propto t^{-0.29}$ and $L \propto t^{-1.2}$ (see equations 1 and 2).

4. For ages up to the age of the Galaxy, crystallization and Debye cooling are unlikely to have any significant effect on the evolution.

5. The theoretically inferred parameters of the brown dwarf VB 8B are: $M = 0.04 - 0.08 \, M_\odot$ (see also McCarthy, Probst, and Low 1985), $\rho_c = 500 - 2000 \, \mathrm{g \, cm^{-3}}$, and $T_c = 1.0 - 1.5 \times 10^6$ K (see also NRJ).

REFERENCES

Alexander, D. R. 1975, *Ap. J. Suppl.*, **29**, 363.
Bahcall, J. N. 1984, *Ap. J.*, **287**, 926.
Bahcall, J. N. 1985, in *"Astrophysics of Brown Dwarfs"*, ed. M. Kafatos (Cambridge University Press), in press.
Grossman, A. S., Hays, D., & Graboske, H. C., Jr. 1974, *Astr. Ap.*, **30**, 95.
Gautier, D., & Owen, T. 1983, *Nature*, **302**, 215.
Hayashi, C., & Nakano, T. 1963, *Prog. Theor. Phys.*, **30**, 460.
Hubbard, W. B. 1973, *Ap. J.*, **182**, L35.
Hubbard, W. B., & Lampe, M. 1969, *Ap. J. Suppl.*, **18**, 297.
Iben, I. Jr. 1975, *Ap. J.*, **196**, 525.
Kumar, S. S. 1963, *Ap. J.*, **137**, 1121.
McCarthy, D. W. Jr., Probst, R. G., & Low, F. J. 1985, *Ap. J.*, **290**, L9.
Lunine, J. I. 1985, in *"Astrophysics of Brown Dwarfs"*, ed. M. Kafatos (Cambridge University Press), in press.
Nelson, L. A. 1985, unpublished results.
Nelson, L. A., Rappaport, S. A., & Joss, P. C. 1985a, *Nature*, **316**, 42 (NRJ).
Nelson, L. A., Rappaport, S. A., & Joss, P. C. 1985b, *Ap. J.*, to be submitted.
Rappaport, S., & Joss, P. C. 1984, *Ap. J.*, **283**, 232.
Rappaport, S., Joss, P. C., & Webbink, R. F. 1982, *Ap. J.*, **254**, 616 (RJW).
Rappaport, S., Verbunt, F., & Joss, P. C. 1983, *Astrophys. J.*, **275**, 713.
Slattery, W. L., Doolen, G. D., & DeWitt, H. E. 1980, *Phys. Rev.*, **A21**, 2087.
Stevenson, D. J. 1978, *Proc. Astr. Soc. Austr.*, **3**, 227.
Stevenson, D. J. 1985, in *"Astrophysics of Brown Dwarfs"*, ed. M. Kafatos (Cambridge University Press), in press.
Tarter, J. 1975, Ph.D. Thesis, University of California, Berkeley.
Zapolsky, H. S., and Salpeter, E. E. 1969, *Ap. J.*, **158**, 809.

This work was supported in part by the National Science Foundation under grant AST-8419834 and by the National Aeronautics and Space Administration under grants NSG-7643 and NGL-22-009-638.

DISCUSSION

Campbell: Since these objects are completely convective, and convection and rotation in more massive stars leads to significant magnetic fields generated by the dynamo mechanism, is it conceivable that such fields in brown dwarfs could significantly affect their structure?

Nelson: It has been conjectured that *fully* convective stars (i.e., stars with no radiative or conductive cores) do not have an operative magnetic dynamo mechanism. However, the final word on this subject is not yet in.

Stevenson: In a convective dynamo, the field will not grow larger than the value given by the Coriolis and Lorentz forces. For a 10 hour rotation period and convective velocities given by mixing length theory, this implies a field strength of approximately one kilogauss. This is much too small to affect the gross structure (i.e., $B^2/8\pi \ll$ internal pressure). The resulting internal field could cause these bodies to be strong radio emitters, however.

Liebert: To what degree did you just use the Debye-Hückel approximation in the regime in which Coulomb effects on the equation of state are important?

Nelson: The Debye-Hückel approximation was used to describe the Coulomb interactions among constituent particles in the limit of a weakly coupled plasma, i.e., for values of the plasma parameter $\Gamma \lesssim 0.3$. For more strongly coupled plasmas ($\Gamma \gtrsim 0.3$), we used the formulation of Slattery, Doolen, and DeWitt (1980) to determine the Coulomb contribution to the equation of state. This approach results in very small errors to the total pressure and specific entropy for the range of temperatures and densities of interest.

D'Antona: How did you take nuclear screening effects into account?

Nelson: We combined the weak screening correction given by Salpeter and Van Horn (1969, *Ap. J.*, **155**, 183) with the strong screening correction given by Itoh *et al.* (1979, *Ap. J.*, **234**, 1079), by use of the prescription suggested by Salpeter and Van Horn.

D'Antona: Do your calculations of the nuclear energy generation rates take into account non-equilibrium phenomena?

Nelson: We have specifically taken into account the breakdown of He^3 nuclear quasi-equilibrium. Nuclear energy release from the burning of He^3 should be relatively unimportant for stars with masses near or below the end of the hydrogen-burning main sequence.

EVOLUTION OF SUBSTELLAR "BROWN" DWARFS
AND THE EVOLUTIONARY STATUS OF VB8B

Guy S. Stringfellow
Lick Observatory, Board of Studies in Astronomy and Astrophysics
University of California, Santa Cruz, California, 95064

Abstract - Evolutionary results for substellar "brown" dwarfs (SBD; M < 0.08 $M_\odot$) and stars near the limiting main-sequence mass are presented and comparisons between the theoretical tracks and VB8B, the recently discovered companion to the star Van Biesbroeck 8, are made. The standard set of stellar structure equations are solved, the physics of which includes degeneracy, screened deuterium burning and the non-equilibrium proton-proton chain, pressure ionization and dissociation, and both molecular and grain opacities. Calculations for both high and low grain opacities, κ_g, are performed in order to assess the impact of this quite uncertain piece of physics. We find all masses < 0.08 $M_\odot$ evolve to $10^{-4} L_\odot$ in $\leq 10^9$ yrs. Both our 0.08 $M_\odot$ and 0.075 $M_\odot$ become main-sequence (MS) stars after 5×10^9 yrs but with luminosities substantially below $10^{-4} L_\odot$. The best fit to VB8B is 0.06 $M_\odot \pm 0.02 M_\odot$, but <0.08 $M_\odot$. Depending upon its *very precise mass*, VB8B can be either a SBD with an age 0.3-3×10^9 yrs, or a MS star (0.075-0.078 $M_\odot$) with an age anywhere from 2×10^9 yrs to a Hubble time. The cooling time for SBD's increase considerably below $10^{-5} L_\odot$, especially for the higher masses. A reduction in κ_g by a factor of 10 generates relatively small deviations in the late evolution of SBD's, the most significant being an increase in the age at a given L by up to a factor of 2 or more.

INTRODUCTION

The recent discovery of a very cool and subluminous companion, VB8B, to the star VB8 (McCarthy, Probst, and Low 1985) stimulated in part the work presented here. Could this object in fact be a low mass MS star, a planet in the early stages of formation, or perhaps not quite either? Provided the observed properties of VB8B hold, the ramifications of any of these possibilites for its identification could be highly significant. We assume VB8B is a bound companion to VB8. If VB8B were a planet forming by accretion within a solar-type nebula, the deduced age would be $\leq 10^7$ yrs for VB8B *and* the primary; therefore, we pursue the other two possible explanations here. Furthermore, it is possible that subluminous, low mass "stars" could be the primary constituents of the dark matter. Pre-existing calculations have concentrated on the minimum MS mass, which was found to lie at ~ 0.08 $M_\odot$ with luminosity $\sim 10^{-4} L_\odot$ (Grossman, *et al* 1970), and masses less than this were not carried to very low luminosities. This study presents detailed evolutionary calculations for stars near and below the minimum MS mass. Hence, SBD specifically refers to stars below the critical minimum mass required to achieve quasi-static hydrogen burning but which have formed according to the usual processes of star formation (*cf* Black 1985; Boss 1985). Unlike previous calculations, these evolutionary sequences are followed to luminosities $\leq 10^{-5} L_\odot$. Furthermore, we hope to clarify which of the many components of the physics is truly important and just how sensitive the results are to uncertainties in the physics.

CALCULATIONS AND RESULTS

The stellar structure equations, including deuterium burning, nonequilibrium proton-proton cycle, and the mixing length treatment for convection, with $\ell/H_p = 1.5$, are solved for nonrotating protostars contracting at constant mass, beginning at the top of their Hayashi tracks. These equations are integrated out to optical depth unity. Reaction rates were obtained from Fowler, *et al* (1975) and Harris, *et al* (1983), and screening (weak, intermediate, or strong) applied according to the theory given by Graboske, *et al* (1973). Contributions to the Rosseland mean opacity include molecules (Alexander, *et al* 1983), and grains comprised of ices, carbon, and silicates both with and without iron (Pollack, *et al* 1985), in addition to the stellar opacities of Cox and Stewart (1970).

The electron gas has been calculated using the technique of Eggleton, *et al* (1973). However, their method of accounting for pressure ionization simply does not work; for temperatures $T < 10^6$ °K, the gas recombines as the density increases, or, for very low T ($< 10^4$ °K) never becomes (fully) ionized at all. We instead account for this effect by making a modification to the Saha equation which emulates the results obtained by more detailed and complex treatments (*cf* Graboske, *et al* 1969). The dissociation of H_2 and the various ionization states of atomic hydrogen, helium, and residual metals of solar composition are thus followed. Corrections to the pressure resulting from other nonideal effects (such as Coulomb interactions) have not yet been included. For the following reasons, however, we deem such modifications as relatively unimportant in the present context: our models, over most of their evolution, are completely convective and adiabatic; degeneracy results in the electron gas behaving more like an ideal gas; and our results compare very favorably with the results of D'Antona (1985), which include some of these effects.

Our main results are shown in Figures 1-3. The surface temperature and luminosity evolution for masses of 0.1 $M_\odot$, 0.075 $M_\odot$, 0.06 $M_\odot$, and 0.04 $M_\odot$ are presented in Figure 1. Empirically derived surface properties of VB8B (McCarthy, *et al* 1985), and VB8 and VB10 (Reid and Gilmore, 1984) are shown for reference. Figure 2 gives the evolution of the central density and temperature. All of the masses go through a short deuterium burning phase of a few million years, but because of the currently low deuterium abundance used, never become fully stabilized on a deuterium MS (see also Fig. 3).

Figure 1   Evolutionary tracks in the H-R diagram calculated with a hydrogen mass fraction X = 0.70, deuterium $X_D = 10^{-5}$, and Z = 0.02 for the metals. Only ages for 0.10 $M_\odot$ (dashed curve) and 0.06 $M_\odot$ (solid curve) are shown in the figure (see Table 1 for additional masses). The errors associated with the observations of VB8B are also indicated. Squares show the region of deuterium burning. The zero-age MS has been reached for the 0.075 $M_\odot$ (dotted curve) and 0.10 $M_\odot$ stars; the unfilled diamond indicates the MS for our 0.08 $M_\odot$ star (track not shown). The 0.06 $M_\odot$ and 0.04 $M_\odot$ (dash-dot curve) are true SBD's, having insufficient nuclear energy production necessary for stabilization. The low κ_g track for the 0.06 $M_\odot$ SBD (not shown) evolves below and to the left of the high κ_g track shown here.

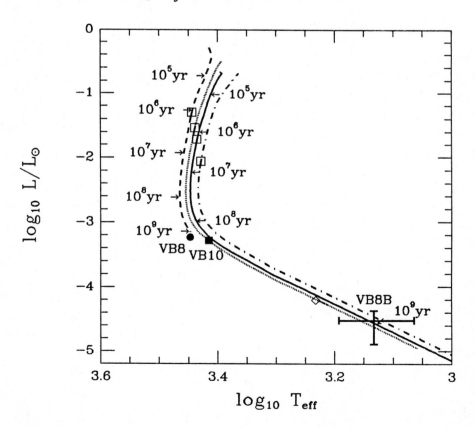

Figure 2   Evolution of the *central* density and temperature for the same tracks shown in Figure 1. All symbols and curves have the same meaning as in Figure 1. All masses ≤ 0.08 $M_\odot$ evolve onto their degenerate cooling tracks and those < 0.075 $M_\odot$ continue cooling as SBD's. The 0.075 $M_\odot$ star actually reaches the zero-age MS prior to attaining its limiting radius (see also Figure 3).

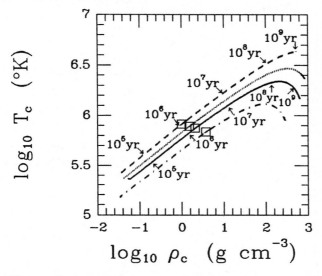

Figure 3   The evolution of the *surface* radius as a function of time. The sequences and squares have been defined in Figure 1. Note that the 0.04 $M_\odot$ SBD achieves a larger limiting radius than the 0.06 $M_\odot$ SBD.

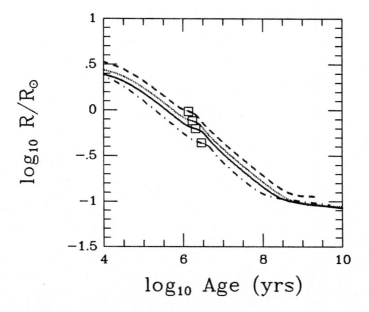

Table 1 Properties of models at equal intervals of luminosity (high κ_g).

$\log_{10} L/L_\odot$	Age (yr)	$\log_{10} \rho_c$	$\log_{10} T_c$	$\log_{10} R/R_\odot$	$\log_{10} T_{eff}$
		0.100 $M_\odot$			
-1.50	2.846×10^6	0.307	6.013	-0.1211	3.4488
-2.00	1.263×10^7	1.080	6.253	-0.3919	3.4592
-2.50	6.766×10^7	1.846	6.471	-0.6531	3.4648
-3.00	3.778×10^8	2.551	6.619	-0.8873	3.4569
		0.075 $M_\odot$			
-1.50	1.277×10^6	0.130	5.876	-0.0993	3.4379
-2.00	7.400×10^6	0.893	6.111	-0.3717	3.4491
-2.50	3.695×10^7	1.647	6.321	-0.6330	3.4548
-3.00	1.753×10^8	2.337	6.458	-0.8630	3.4448
-3.50	4.864×10^8	2.690	6.457	-0.9826	3.3796
-4.00	9.528×10^8	2.809	6.426	-1.0243	3.2754
-4.50	2.399×10^9	2.864	6.401	-1.0441	3.1603
		0.060 $M_\odot$			
-1.50	6.671×10^5	-0.005	5.771	-0.0816	3.4291
-2.00	5.231×10^6	0.748	6.002	-0.3561	3.4413
-2.50	2.343×10^7	1.492	6.206	-0.6177	3.4471
-3.00	1.065×10^8	2.166	6.332	-0.8440	3.4353
-3.50	2.734×10^8	2.502	6.328	-0.9587	3.3676
-4.00	4.627×10^8	2.603	6.302	-0.9945	3.2605
-4.50	8.784×10^8	2.661	6.278	-1.0160	3.1463
-5.00	2.276×10^9	2.721	6.243	-1.0388	3.0327
-5.50	7.199×10^9	2.784	6.189	-1.0647	2.9206
		0.040 $M_\odot$			
-1.50	2.808×10^5	-0.227	5.590	-0.0526	3.4146
-2.00	1.845×10^6	0.488	5.811	-0.3277	3.4271
-2.50	1.112×10^7	1.219	6.004	-0.5904	3.4335
-3.00	4.493×10^7	1.853	6.110	-0.8119	3.4192
-3.50	1.047×10^8	2.147	6.103	-0.9149	3.3457
-4.00	1.704×10^8	2.238	6.082	-0.9481	3.2373
-4.50	3.305×10^8	2.297	6.061	-0.9707	3.1236
-5.00	8.760×10^8	2.359	6.029	-0.9956	3.0111
-5.50	2.754×10^9	2.425	5.983	-1.0237	2.9001

Our 0.1 $M_\odot$ star is in good agreement with those of Grossman, *et al* (1970), VandenBerg, *et al* (1983), and D'Antona (1985), reaching the MS in 2.25×10^9 yrs with $\log_{10}L/L_\odot = -3.151$ and $\log_{10}T_{eff} = 3.450$ (varying ℓ/H_p results in the same MS values). In our definition of the (hydrogen burning) MS, the zero-age is attained when T_{eff} and L reach their minimum values and begin to increase. Note that if an equilibrium proton-proton chain is used, the overestimated energy produced would increase the MS position to higher values. Masses $\leq 0.08M_\odot$ reach a maximum T_c and begin to cool, as shown in Figure 2, due to the onset of degeneracy. The 0.075 $M_\odot$ star also stabilizes on the MS with $\log_{10}L/L_\odot = -4.946$, $\log_{10}T_{eff} = 3.059$ and an age of 5.4×10^{10} yrs; an extremely low L and a *very long time!* VB8B, therefore, could be a MS star, lying within the observational errors, between 0.075 and 0.078 $M_\odot$. Screening of the nuclear reaction rates enhances the energy production, increasing the stellar luminosity by a factor of ~2 throughout the nuclear evolution. If screening is turned off, the 0.075 $M_\odot$ does not become fully stabilized ($L_{nuc}/L \simeq 70\%$ maximum) and proceeds to cool as a SBD, but over very long times. Table 1 gives the age and central and surface properties at equal intervals of luminosity for the masses shown in Figures 1-3.

During this workshop, much concern has been expressed over the uncertainty in the opacities used in the calculations. It turns out that the difference between Alexander's (1975) and Alexander, *et al* (1983) *molecular* opacities is relatively small over the range of atmospheric densities and temperatures realized in these calculations; $\rho_c \geq 10^{-7}$g cm^{-3} and $T \geq 2000$ °K. However, substantial variations prevail in κ_g and, to test the sensitivity of our results to this variation, the 0.06 $M_\odot$ SBD has been rerun with κ_g reduced by ~1 order of magnitude below 2000 °K. Because of the T dependence, deviations from our standard high κ_g become evident only when $L \leq 10^{-4}L_\odot$: at $10^{-5}L_\odot$, T_{eff} is increased by only 71 °K and the age is increased by a factor of 2; at 1140 °K, L is reduced by 25%,

Table 2   Surface and central properties and ages from our evolutionary calculations at the empirically derived temperature of VB8B. The surface density and opacity are denoted by ρ_s and κ_s, respectively.

M/M$_\odot$	Age (yr)	$\log_{10}$ L/L$_\odot$	$\log_{10}$ R/R$_\odot$	$\log_{10}$ ρ_s	$\log_{10}$ κ_s
		VB8B : T_{eff} = 1360 °K			
0.075	3.75×10^9	−4.618	−1.049	−4.390	−0.877
0.075*	2.28×10^9	−4.618	−1.050	−4.390	−0.877
0.060	1.55×10^9	−4.607	−1.044	−3.104	−2.271
0.060	9.21×10^8	−4.554	−1.017	−4.523	−0.905
0.040	3.10×10^8	−4.457	−0.969	−4.748	−0.953

*Evolution with unscreened reaction rates

accompanied by a factor 2.6 increase in the age. The major difference appears as an increase in the age, a smaller radius, and a compensating increase in the atmospheric pressure. This behavior occurs because the SBD is still contracting while cooling. Hence, the release of gravitational energy, E_g, is still important. A larger opacity traps more energy and increases the radius. Since $|E_g| \propto R^{-1}$, and the time is given roughly by $|E_g|/L$, we see that an increase in R yields shorter evolutionary times to a specified L. This is also true for a contracting protostar. Thus, use of analytic relations based on strict cooling curves is inappropriate for describing the evolution of SBD's prior to reaching their limiting radius, which occurs below $10^{-5}L_\odot$. None of the SBD's have quite reached their limiting radius at the end of the calculation, as is evident from Table 1 and Figure 3.

CONCLUSIONS

Comparisons between our calculations and the empirically determined temperature and derived luminosity of VB8B are given in Table 2. In order to have such a low L (within the estimated errors), VB8B must be less than 0.08 $M_\odot$. Similarly, consistency with the age spread given by the mass range for VB8 ($\sim$0.08-0.11 $M_\odot$) requires VB8B be $\geq$0.04 $M_\odot$. The best fit yields 0.06 $M_\odot \pm 0.02M_\odot$ and an age of 0.9-1.6×10^9 yrs, depending on κ_g. However, details concerning the equation of state must also be responsible for small deviations in the age: our SBD's evolve quicker during their early evolution when compared with D'Antona's (1985) results. At the luminosity of VB8B, D'Antona obtains an age of about 2×10^9 yrs, in reasonable agreement with our low opacity case. SBD's with mass < 0.04 $M_\odot$ evolve below $10^{-5}L_\odot$ in less than 3.5×10^8 yrs and are not expected to be observed outside of regions of active star formation. A greater number of SBD's over a wider mass range may be found between $10^{-6} \leq L/L_\odot \leq 10^{-5}$ since their rate of evolution slows considerably through this interval.

I would like to thank Drs Peter Bodenheimer and David Black for many stimulating discussions and suggestions. This work was supported in part by NASA grant NGT 05 061 800 and performed under the auspices of a special NASA Astrophysics Theory program which supports a joint Center for Star Formation Studies at NASA-Ames Research Center, U.C. Berkeley, and U.C. Santa Cruz.

References

Alexander, D. R. 1975, Ap. J. Suppl. *29*, 363.
Alexander, D. R., Johnson, H. R., and Rypma, R. L. 1983, Ap. J. *272*, 773.
Black, D. C. 1985, this workshop.
Boss, A. P. 1985, this workshop.
Cox, A. N., and Stewart, J. N. 1970, Ap. J. Suppl. *19*, 243.
D'Antona, F. 1985, this workshop.
Eggleton, P. P., Faulkner, J., Flannery, B. P. 1973, Astr. Ap. *23*, 325.

Fowler, W. A., Caughlan, G. R., Zimmerman, B. A. 1975, Ann. Rev. Astr. Ap. *13*, 69.

Graboske, H. C., DeWitt, H. E., Grossman, A. S., and Cooper, M. S. 1973, Ap. J. *181*, 457.

Graboske, H. C., Harwood, D. J., and Rogers, F. J. 1969, Phys. Rev. *186*, 210.

Grossman, A. S., Mutschlecner, J. P., and Pauls, T. A. 1970, Ap. J. *162*, 613.

Harris, M. J., Fowler, W. A., Caughlan, G. R., and Zimmerman, B. A. 1983, Ann. Rev. Astr. Ap. *21*, 165.

McCarthy, D. W., Probst, R. G., Low, F. J. 1985, Ap. J. Lettr. *290*, L9.

Pollack, J. B., McKay, C. P., and Christofferson, B. M. 1985, preprint.

Reid, N., and Gilmore, G. 1984, M.N.R.A.S. *206*, 19.

VandenBerg, D. A., Hartwick, F. D. A., Dawson, P., and Alexander, D. R. 1983, Ap. J. *266*, 747.

DISCUSSION

Shipman: A question for Bob Harrington. Many speakers have mentioned a mass of 0.05 $M_\odot$ for VB8B. Yesterday you quoted 0.02 $M_\odot$ "$\pm$". How comfortable are you with 0.05 $M_\odot$ in terms of the astrometry?

Harrington: I'm perfectly comfortable with 0.05 $M_\odot$. It's the lower masses, $<$ 0.01 $M_\odot$, which present problems astrometrically. You can make the mass higher by adjusting the angle of inclination.

Probst: Your own and other evolutionary models show the ZAMS locus of minimum mass stars to be $\log_{10} L/L_\odot \simeq -3.3$ and $\log_{10} T_{eff} \simeq 3.44$, in rough agreement with VB10. D'Antona and Mazzetelli's models put it much lower (-4.4, 3.2) with a considerable overlap in (L, T_{eff}) of old stars and young brown dwarfs. Where do the very different results come from and which is to be believed? This is critical for interpreting the results of BD searches in terms of populations (number densities, etc.).

Stringfellow: The main-sequence can extend down to very low L and T; in fact, down to L $\simeq 10^{-5} L_\odot$ (only the 0.06 and 0.10 $M_\odot$ tracks were shown during the workshop). However, this faint extension occurs only over a very narrow range in stellar mass: $\sim$0.075-0.08 $M_\odot$. The precise values depend somewhat on the physics utilized. This is a consequence of the non-equilibrium p-p reaction.

D'Antona: Neece (1984, Ap.J. *277*, 738) has shown that the use of an approximated equation of state produces a larger luminosity for a given mass with respect to a more correct EOS. This seems to be happening in your 0.1 $M_\odot$ model too, in spite of the fact that you use very large opacities, which tend to decrease the model luminosity. So your agreement with Grossman's models is probably fortuitous. I must stress again that the low luminosity of my 0.085 $M_\odot$ model depends on the molecular opacities of Alexander, *et al* (1983), but not on the grains.

THE POSITION OF BROWN DWARFS ON THE UNIVERSAL DIAGRAMS

Minas Kafatos
Department of Physics, George Mason University, Fairfax, VA

Abstract. A series of universal diagrams have been constructed.
Various physical quantities have been plotted against each
other on these diagrams, including mass, size, luminous
output, entropy radiated away and angular momentum. I have put
as many objects on these diagrams as I could including brown
dwarfs. The diagrams show continuity among various classes
of objects and brown dwarfs fall naturally between planets
and stars. These diagrams can even be used to roughly predict
where objects should lie on them and, therefore, deduce physical properties of fairly unknown objects such as brown
dwarfs. Using the entropy-mass universal diagram presented
here I conclude that brown dwarfs could make a substantial
contribution to the dark matter density in the solar vicinity.
Moreover, the spin angular momentum of VB8 B could be quite
high and detectable.

INTRODUCTION

A number of universal diagrams (UD) can be constructed. These
are diagrams where as many classes of objects in the universe as possible
are shown. The quantities plotted are mass, size, luminosity, surface
temperature, angular momentum and entropy change of the universe resulting
from the radiation of the objects in it. Mass is equivalent to energy
and for massless particles (e.g. photons) the quantity that should be
plotted is E/c^2, where E is the energy and c is the speed of light. One
such diagram has been published before by the author, namely the luminosity-mass UD (Kafatos 1985). Luminosity is an important physical quanity
since we only know directly the presence of luminous matter in the universe. Surface temperature characterizes the type of radiation put out
by an object. Entropy change of the universe due to radiation by an object is an important thermodynamic quantity. Mass (or energy equivalent
mass) is, of course, important quantity.

In the present work I present two diagrams: Figure 1 is the luminosity-
temperature diagram. Figure 2 is the entropy radiated by the shown objects over their lifetimes versus their masses. I have chosen here to
present only the macroscopic portions of these diagrams, i.e. the objects
shown are generally larger than comets.

Even though the particular details of these diagrams are not unique-
somebody else would probably draw slightly different objects in the
diagrams-their overall appearance does not change. Since the quantities
plotted vary by many orders of magnitude, the UDs are extremely compact
and even order of magnitude uncertainties do not change their overall appearance.

THE UNIVERSAL DIAGRAMS
1. Black Holes
Black holes are the simplest objects in the universe. As we
see from Figures 1 and 2 black holes divide these diagrams into two
regions. Primordial mini-black holes would be radiating their energy away
(Hawking 1976) and would have decayed away within the age of the universe
if their initial mass were less than 5×10^{14} gr (Page 1976a). Rotating
black holes of this order of magnitude would spin down rapidly to a nonrotating type before their mass was radiated away (Page 1976b). For a nonrotating black hole the luminosity, temperature, entropy and lifetime
are given, respectively (Hawking 1976, Page 1976a):

$$L \sim 2\hbar c^6 G^{-2} M^{-2} \qquad\qquad S = 4\pi k G M^2 / c\hbar$$

$$T = \hbar c^3 / (8\pi k G M) \qquad\qquad \tau \sim 10^{-28} M^3 \text{ sec} \qquad\qquad (1)$$

where $\hbar$ is Planck's constant divided by 2π and k is Boltzmann's constant
(cf. Kafatos 1985).

2. Brown Dwarfs
In Figure 1, all objects are found in the upper part of the
diagram above the dashed line where mini-black holes would be found. The
familiar H-R diagram occupies a small portion of the UD. The approximate
location of brown dwarfs is shown below the main sequence and the position
of VB8 B is also shown. I have assumed for VB8 B that $L = 3 \times 10^{-5} L_\odot$, and
$T = 1360$ K (McCarthy, Probst and Low 1985). Note that brown dwarfs are
found halfway (logarithmically) between the sun and Jupiter. More surprising, however, is the tendency of solar system objects to form a monotonic relationship extending from comets to brown dwarfs. Just by rough
inspection of this diagram and continuing solar system objects to the
M-dwarfs, one would expect brown dwarfs to be $\sim$ an order of magnitude
fainter than VB8 B and with an effective temperature ≤ 1000 K. VB8 B may
indeed be a bright brown dwarf and its detection would not be, therefore,
unexpected. Two man-made objects are also shown for comparison, a one
megaton H-bomb explosion and a hypothetical starship weihing $\sim 10^{11}$ gr
and travelling at speeds $\sim 0.1c - 0.9c$.

In Figure 2 I show the entropy radiated by objects in the universe over
their lifetime versus their mass. The entropy radiated is defined as:

$$S = \int_\tau L \, dt/T \qquad\qquad (2)$$

where τ is the lifetime of the object. The total entropy of an object
is more difficult to define (it assumes a simple expression only in

Figure 1: Universal diagram of luminosity versus effective
temperature. The dashed line indicates the locus of primordial mini-black holes. Brown dwarfs occupy the region of the
diagram between solar system objects and the M-dwarfs.

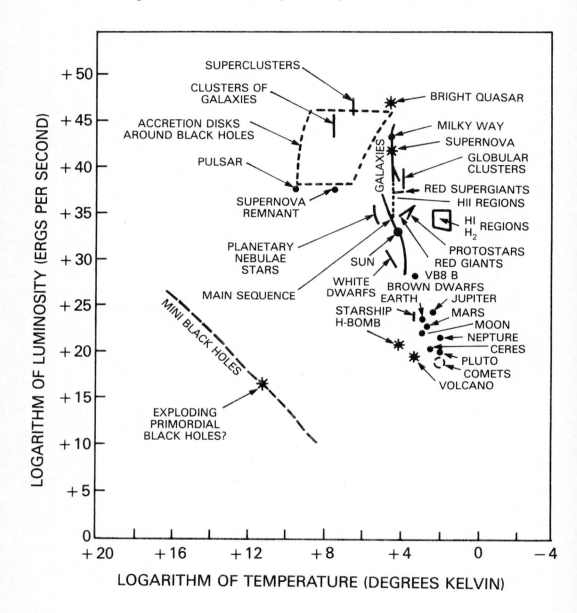

certain cases, e.g. for ideal gases). In any case, if we live in a closed
universe and the universe will go through a reversed cycle, the entropy
radiated away takes on a special meaning because of it the expanding
cycle differs from the contracting cycle.

To a surprising degree of accuracy most objects on this UD follow the
simple relationship $S \sim 2 \times 10^{11}$ M erg K^{-1}: for a given mass no object

Figure 2: Universal diagram of entropy radiated by objects in
the universe over their lifetime versus their mass. The dashed
line indicates black holes except that in contrast to Figure
1 these black holes are macroscopic. For VB8 B see the text
for the parameters assumed. Notice that objects seem to follow
the relation $S \sim 2 \times 10^{11}$ M erg K^{-1} to a surprising degree of
accuracy.

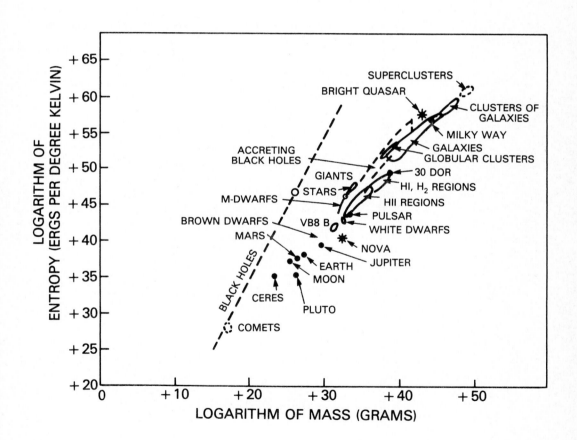

deviates in entropy more than 3 orders of magnitude either way even though the objects plotted vary by 36 orders of magnitude. For a given entropy no object deviates in mass from the above relationship more than 2 orders of magnitude even though the objects plotted vary by 34 orders of magnitude. Note that for black holes (see equations 1) the entropy radiated is proportional to M^2. Why would most objects in the universe obey a linear relationship between entropy and mass is unknown to the author.

Focusing more on the region between Jupiter and VB8 B we expect that brown dwarfs of mass $M \sim 10^{31}$gr ($M \sim 0.005\ M_\odot$) would be radiating $\sim 10^{41}$erg K^{-1} over their lifetime. Using $L \sim 10^{-7}L_\odot$ for such objects from the luminosity-mass UD (Kafatos 1985) and a luminosity-temperature relationship $T \sim 1.8 \times 10^4 (L/L_\odot)^{1/4}$ (cf. Staller and deJong 1981) we find that such objects would be radiating at those luminosities $\sim 2 \times 10^9$yr in rough agreement with the theoretical models of Stevenson (1986).

CONCLUSIONS
The surprising linear relationship between entropy radiated and mass gives a reason to more closely examine how entropy can be used to, perhaps, try to predict other properties of brown dwarfs.

For a given spectral class of stars the total entropy radiated over the main sequence lifetime of the stars is:

$$S_{tot} = \int_{M_1}^{M_2} (\xi d\log M\ L\ t_{MS})/T \tag{3}$$

where ξ is the logarithmic initial mass function (IMF) given by Miller and Scalo (1979) and Garmany, Conti and Chiosi (1982). For M-dwarfs I used the Salpeter function (D'Antona 1986) where the minimum mass for red dwarfs was assumed to be 0.5 $M_\odot$. In (3) I assumed that the stellar birth rate has remained constant over the lifetime of the galaxy. The results of estimating (3) are given in Table 1.

Table 1.
Total Entropy Radiated as a Function of Spectral Class

Spectral Class	Mass Range ($M_\odot$)	Entropy (erg $K^{-1}pc^{-2}$)
O	$\gtrsim 30$	3.6×10^{47}
B	3.5 – 30	10^{48}
A	1.8 – 3.5	2.6×10^{48}
F	1.1 – 1.8	3.7×10^{48}
G	0.8 – 1.1	2.8×10^{48}
K	0.5 – 0.8	1.8×10^{48}
M	0.08 – 0.5	10^{48}
Brown Dwarfs*	0.01 – 0.08	$\sim 10^{47} - 10^{48}$

*Estimated

Extrapolating the radiated entropy to the brown dwarf mass range
(M $\lesssim$ 0.08 M$_\odot$), the entropy radiated by brown dwarfs is estimated to be
$\sim 10^{47}$ - 10^{48} erg K^{-1}pc^{-2} (see Table 1). The greatest contribution to
the total entropy radiated is provided by brown dwarfs near the upper
mass limit of 0.08 M$_\odot$. Taking S$\sim$ 2x10^{43}erg K^{-1}for such objects computed
by using Stevenson's results (Stevenson 1986) in relation (2), we find
a surface density of brown dwarfs in the range 5x10^3 - 5x10^4 pc^{-2}. This
is approximately 70 - 700 the value for red dwarfs. Conservatively, the
nearest brown dwarf would be $\sim$0.4 pc away. It is interesting to note
that D'Antona (1986) reached similar conclusions following an entirely
different argument. We conclude that based on this order of magnitude
argument presented here that brown dwarfs could make a significant
contribution to the local mass density of 0.1 M$_\odot$ pc^{-3} (Bahcall 1984).

The UDs cannot, obviously, be used to obtain accurate physical parameters
of objects (say, better than an order of magnitude). They serve, however,
at least one important purpose: The position of poorly known classes of
objects in these diagrams can be roughly estimated. One such class of
objects is, of course, brown dwarfs. To illustrate the usefulness of
the UDs we try to estimate the angular momentum of the VB8 double star
system and the spin of VB8 B--by using the angular momentum-mass UD
(Kafatos, in preparation). Assuming a mass of $\sim$ 0.1 M$_\odot$ for VB8 A
(McCarthy, Probst and Low 1985) we find from this UD that the angular
momentum of VB 8 is $\sim$ 5x10^{51} - 2x10^{52} erg s. This is found by extending
the angular momentum line for double stars to the total mass point for
VB8, $\leq$ 0.18 M$_\odot$. The value found is close to the angular momentum of
VB8 B around VB8 A computed by assuming that the mass of VB8 B is in the
approximate range 0.03 - 0.08 M$_\odot$ and that the distance between them
is $\sim$ 6.5 A.U. (McCarthy, Probst and Low 1985). This value computed
by using the above values is $\sim$ 1.5x10^{51}- 4x10^{51}erg s. Extending the spin
line in the angular momentum-mass UD from planetary objects through the
brown dwarf mass range to stars, we find a spin angular momentum for
VB8 B in the approximate range 4x10^{48} - 10^{49}erg s for M $\sim$0.03 - 0.08 M$_\odot$,
respectively. The corresponding period of rotation of VB8 B would then
be $\sim$ 25 minutes and independent of mass while the rotational velocity
would be $\sim$ 260 km s^{-1}. This value is close to the break-up velocity
for an object of mass $\sim$ 0.03 M$_\odot$ (if the radius is $\sim$ 0.09 R$_\odot$--McCarthy,
Probst and Low 1985). Obviously the mass of VB8 B cannot be much less
than $\sim$ 0.03 M$_\odot$. The UDs provide again an estimate for the minimum mass
of VB8 B which is in agreement with our ideas about the nature of VB8 B
(cf. D'Antona 1986). As far as I know, no one has given a theoretical
reason for a high spin of VB8 B. At 10 μ the rotational broadening would
be $\sim$ 100 Å and should be detectable with high resolution infrared
spectroscopy (cf. Lunine 1986).

REFERENCES

Allen, C.W. (1964). Astrophysical Quantities. London: University of London.
Bahcall, J.N. (1984). Ap. J., 287, 929.
Carr, B.J. & Rees, M.J. (1979). Nature, 278, 605.
D'Antona, F. (1986). Present volume.
Garmany, C.D., Conti, P.S. & Chiosi, C. (1982). Ap. J., 263, 777.
Hawking, S.W. (1976). Phys. Rev. D, 13, 191.
Kafatos, M. (1985). In The Search for Extraterrestrial Life: Recent
        Developments, IAU Symposium No. 112, ed. M.D. Papagiannis,
        pp. 245 - 249. Dordrecht: D. Reidel.
Lunine, J.I. (1986). Present volume.
McCarthy, D.W., Probst, R.G. & Low, F.J. (1985). Ap. J. (Letters), 290,
        L9.
Miller, G.E. & Scalo, J.M. (1976). Ap. J. Suppl. Series, 41, 513.
Page, D.N. (1976a). Phys. Rev. D, 13, 198.
Page, D.N. (1976b). Phys. Rev. D, 14, 3260.
Staller, R.F.A. & de Jong, T. (1981). A & A, 98, 140.
Stevenson, D.J. (1986). Present volume.

DISCUSSION

V.L. Teplitz: What are the limits on the angular momentum of the universe?

M. Kafatos: The limit comes from the observed isotropy of the micro-
wave background which results in an angular velocity of the
universe less than about 10^{-6} arc sec per century.

THEORETICAL DETERMINATION OF THE MINIMUM PROTOSTELLAR MASS

A. P. Boss
DTM, Carnegie Institution of Washington
5241 Broad Branch Road NW, Washington DC 20015, U.S.A.

Abstract. The minimum mass of protostars formed from the collapse and fragmentation of Population I interstellar clouds is found to be about $0.02M_\odot$. This lower bound is determined by the mass of the fragments of the least massive cloud that is able to fragment into a binary system. The calculations rigorously treat the hydrodynamical collapse in three spatial dimensions, including self-gravity, radiative transfer in the Eddington approximation, and realistic equations of state.

INTRODUCTION

Knowledge of the minimum protostellar mass is important to the study of brown dwarfs for several reasons. First, models of the thermal history of brown dwarfs can be constructed for arbitrarily low masses and then compared to observations of objects such as Van Biesbroeck 8B = VB8B (Harrington et al. 1983; McCarthy et al. 1985). The observations can be matched by models of brown dwarfs (e.g., Nelson et al. 1985) which are either relatively massive ($\sim 0.08M_\odot$) and old ($\sim 10^9 years$) or less massive ($\sim 0.01M_\odot$) and young ($\sim 10^7 years$). Obviously a firm lower bound on the mass of protostars rules out thermal history models for objects with masses below the minimum mass. Second, perhaps the most exciting facet of brown dwarfs is the possibility of accounting for the unseen mass in the solar neighborhood through a large population of brown dwarfs (Bahcall et al. 1985). Because the initial mass function is poorly defined for low mass stars, much less for brown dwarfs, estimates of the total amount of mass contributed by brown dwarfs must rely at present on uncertain extrapolations of the initial mass function through the brown dwarf mass range. The total integrated mass is then dependent on the lower limit given by the minimum protostellar mass. Finally, if the minimum protostellar mass should turn out to be larger than the best estimates of the mass of objects such as VB8B, this inconsistency would require the revision of either the theory that produced the lower bound or else of the observation and its interpretation.

In this paper we will assume that both stars (hydrogen burning) and brown dwarfs (non-hydrogen burning) result from the formation of protostars, which are themselves formed primarily through the collapse and fragmentation of rotating, dense, interstellar clouds. One or more stages of hierarchical fragmentation then appear to be necessary in order to account for the formation of binary systems and multiple systems consisting of pairs of binaries.

Previous estimates of the minimum protostellar mass were based on analytical models of opacity-limited hierarchical fragmentation (Low & Lynden-Bell 1976; Rees 1976; Silk 1977). Because the initial phases of the gravitational collapse of interstellar clouds occur at approximately constant temperature, the Jeans mass $M_J \propto T^{3/2} \rho^{-1/2}$ decreases as the density increases, which was hypothesized to lead to fragmentation of progressively smaller masses (Hoyle 1953). Once the cloud (primarily H_2) becomes optically thick and begins to heat up, $T \propto \rho^{2/5}$, and so the Jeans mass ($M_J \propto \rho^{1/10}$) thereafter increases. Thus it was argued that the Jeans mass reaches a minimum value when the cloud becomes optically thick, leading to estimates of the minimum protostellar mass in the range of $0.003 - 0.007 M_\odot$. However, these estimates cannot be considered rigorous, for two reasons. First, unit optical depth does not ensure that heating of the cloud is taking place (F. Shu, private communication). For example, in an optically thick cloud where the cooling rate exceeds the heating rate, the temperature will not increase significantly. A proper treatment of the time dependent terms in the energy equation is required. Second, fragmentation is a complicated process involving effects neglected in the Jeans mass derivation, such as nonlinear growth, finite size of the fragmenting cloud, rotation, and nonuniform density, temperature, and velocity fields (see also Tohline 1980).

NUMERICAL TECHNIQUES

In order to improve upon the analytical estimates of the minimum protostellar mass, one needs to solve the fully three dimensional equations of motion: the continuity equation, three momentum equations, the energy equation, the Poisson equation for the gravitational potential, a radiative transfer equation, and equations of state for the pressure, internal energy, and opacity. This has been accomplished numerically, using a computer code developed for the study of the collapse and fragmentation of interstellar clouds. Details of the numerical methods and test cases are given by Boss (1980, 1984).

The code uses modified donor-cell hydrodynamics to solve the equations of motion on a moving, spherical coordinate grid, with the Poisson equation being solved by an expansion in spherical harmonics. Because the equations of motion are solved in conservation law form, the total mass and angular momentum are each conserved throughout the calculations to about four digits accuracy. Radiative transfer is handled by the three dimensional extension of the Eddington approximation (with $f = 1/3\mathbf{I}$), solved by an alternating directions implicit (ADI) method. Rosseland mean opacities are used, including contributions from grains using the routine developed by DeCampli & Cameron (1979).

Test calculations (Boss 1984) have verified the numerical accuracy up to central temperatures of $\sim 2000 K$, which is more than adequate for the physical regime explored in these models (roughly densities of $10^{-18} - 10^{-8} g cm^{-3}$, temperatures of $10 - 1000 K$). The numerical resolution is sufficient ($N_r = 41, N_\theta = 23, N_\phi = 32$) to determine whether or not the $m = 2$ mode grows and fragments the protostar. Each model calculation requires roughly a week of VAX-11/780 CPU time. In order to minimize the computer time, symmetry through the rotation axis is assumed, thereby forcing any binaries that form to have equal masses.

RESULTS

For uniform density and rotation clouds of fixed composition ($X = 0.769, Y = 0.214, Z = 0.017$), a four dimensional parameter space exists. The four parameters which have been varied are the mass ($0.02 - 2M_{\odot}$), the initial temperature (10 K, 40 K), and the initial ratios of thermal ($\alpha = 0.01 - 0.5$) and rotational ($\beta = 0.04 - 0.000004$) energy to the gravitational energy. These two temperatures correspond roughly to cold, dark clouds, and to clouds heated by the prior formation of OB stars, respectively. Significantly higher values of α or β would preclude collapse, while lower values of α are not obtained in hierarchical fragmentation. The minimum value of β is that of a cloud spinning with the orbital angular velocity of the sun about the galactic center, and hence should be a lower bound on the amount of spin angular momentum in a protostellar cloud. Each cloud is given an initial density perturbation ($\delta\rho = 0.1cos(2\phi)$) which is biased toward binary formation.

Three distinct configurations are obtained (Figure 1): binary systems, rapidly rotating single protostars with spiral arms, and slowly rotating, nearly axisymmetric protostars. The results for varied initial α and M but fixed $\beta = 0.04$ and $T = 10K$ are shown in Figure 2, as well as the properties of the fragments for the clouds that form binary systems. The oblique lines in Figure 2 separate permitted initial conditions (above the line) from thermodynamically inaccessible initial conditions. Calculation of over fifty models has shown that fragmentation depends weakly on mass, temperature, and rotation, but strongly on the thermal energy, with fragmentation generally occurring for low mass protostars whenever $\alpha < 0.1$. The smallest mass protostar which can still collapse and fragment has a mass of about $0.1M_{\odot}$, and produces fragments with a mass of about $0.02M_{\odot}$.

Figure 1. Results of models with $\alpha = 0.25, \beta = 0.04, T = 10K$, and $M = 2.0, 0.1$, and $0.02M_{\odot}$ (left, center, right). Box sizes are 340, 34, and 14 AU, respectively.

Figure 2. Results of models as a function of initial conditions (left),
and fragment properties for models which formed binaries (right).

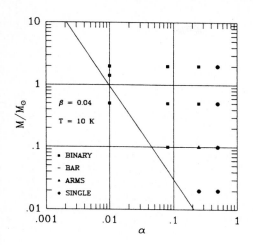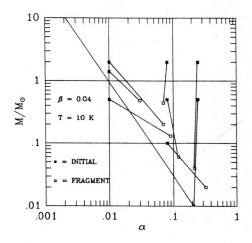

In hierarchical fragmentation starting from relatively massive ($\sim 10^2 - 10^4 M_\odot$) clouds, fragmentation may continue until protostars with masses close to the minimum are obtained. This is evident from Figure 2a, where fragment properties evolve toward decreasing M and α, thereby avoiding the *island of stability* where single protostars form, until the clouds begin to heat, forcing them along the oblique line toward stability at the minimum mass. In this case, one might expect the initial mass function to have a (local) maximum around $0.02 M_\odot$.

CONCLUSIONS

The minimum protostellar mass of $0.02 M_\odot$ applies to regions with dust grain opacity similar to our galactic disk (Population I). The result appears to be a firm lower bound, because other effects, such as magnetic fields, turbulent viscosity, subsequent gas accretion, coalescence with other protostars, centrally condensed initial conditions, less well ordered initial perturbations, and binary orbital decay should all *increase* the minimum mass. Effects which could *decrease* the minimum mass, such as fission during subsequent evolution or increased spatial resolution, have been shown to be unimportant (Durisen *et al.* 1985; also, doubling N_ϕ has little effect on the growth of the $m = 2$ mode). The effects of lowered opacity remain to be explored.

The formation of *brown dwarfs* with $M > 0.02 M_\odot$ is thus consistent with the theory of hierarchical protostellar fragmentation. Less massive objects cannot form from the collapse and fragmentation of interstellar clouds; these objects (*planets*) must form by a different process, such as accumulation in a disk about a protostar.

This research was partially supported by NSF grant AST 83-15645. The calculations were performed on the Carnegie Institution of Washington VAX computers.

REFERENCES

Bahcall, J.N., Hut, P., & Tremaine, S. (1985). Maximum mass of objects that constitute unseen disk material. *Astrophys. J.*, **290**, 15-20.

Boss, A.P. (1980). Protostellar formation in rotating isothermal clouds. I. Numerical methods and tests. *Astrophys. J.*, **236**, 619-27.

Boss, A.P. (1984). Protostellar formation in rotating isothermal clouds. IV. Nonisothermal collapse. *Astrophys. J.*, **277**, 768-82.

DeCampli, W.M. & Cameron, A.G.W. (1979). Structure and evolution of isolated giant gaseous protoplanets. *Icarus*, **38**, 367-91.

Durisen, R.H., Gingold, R.A., Tohline, J.E., & Boss, A.P. (1985). The binary fission hypothesis: a comparision of results from finite-difference and smoothed particle hydrodynamics codes. *Astrophys. J.*, submitted.

Harrington, R.S., Kallarakal, V.V., & Dahn, C.C. (1983). Astrometry of the low-luminosity stars VB8 and VB10. *Astron. J.*, **88**, 1038-9.

Hoyle, F. (1953). On the fragmentation of gas clouds into galaxies and stars. *Astrophys. J.*, **118**, 513-28.

Larson, R.B. (1985). Cloud fragmentation and stellar masses. *Mon. Not. R. Astr. Soc.*, **214**, 379-98.

Low, C. & Lynden-Bell, D. (1976). The minimum Jeans mass or when fragmentation must stop. *Mon. Not. R. Astr. Soc.*, **176**, 367-90.

McCarthy, D.W. Jr., Probst, R.G., & Low, F.J. (1985). Infrared detection of a close cool companion to Van Biesbroeck 8. *Astrophys. J.*, **290**, L9-13.

Nelson, L.A., Rappaport, S.A., & Joss, P.C. (1985). On the nature of the companion to Van Biesbroeck 8. *Nature*, **316**, 42-4.

Rees, M.J. (1976). Opacity-limited hierarchical fragmentation and the masses of protostars. *Mon. Not. R. Astr. Soc.*, **176**, 483-86.

Silk, J. (1977). On the fragmentation of cosmic gas clouds. II. Opacity-limited star formation. *Astrophys. J.*, **214**, 152-60.

Tohline, J.E. (1980). The gravitational fragmentation of primordial gas clouds. *Astrophys. J.*, **239**, 417-27.

DISCUSSION

S. Stahler - Won't uncertainties in your treatment of radiative transfer affect your value of the minimum mass?

A. Boss - Yes, the minimum mass is sensitive to the internal temperature of the gas. The test cases I have run as well as work by others implies that my solution should be reasonably accurate in the desired regime, which is prior to the formation of protostellar photospheres.

B. Tarter - Because all your stars must fragment into equal masses, does this affect the conclusion? Also, what is the opacity dependence of your results?

A. Boss - Because the binary mode tends to dominate fragmentation, I would expect the binary members to have roughly equal masses (and hence little effect on the conclusion) even in a calculation allowing for unequal masses. These calculations are all for fixed opacity. In the next year I hope to study the effects of different opacities.

J. Tarter - Larson's (1985) description of fragmentation concludes that the characteristic stellar mass should be about $0.3 M_\odot$. Why does this result differ from your minimum mass?

A. Boss - Larson's work is based on studying the stability of equilibrium configurations, and comparing the stability criteria to results of dynamic calculations like those presented here. Larson considers that fragmentation occurs only once, in a quasiequilibrium configuration in a relatively diffuse state, whereas my dynamic calculations indicate that several stages of fragmentation may occur, leading to progressively smaller masses.

J. Bahcall - What about other effects that you have not included, such as T-Tauri winds and shock compression from HII regions?

A. Boss - My calculations rigorously apply only to first generation protostars forming in isolation. I expect however that the inclusion of these other effects would increase the minimum mass, for example by adding thermal energy to the protostellar cloud.

POPULATION II BROWN DWARFS AND DARK HALOES

H Zinnecker
Royal Observatory, Blackford Hill,
Edinburgh EH9 3HJ, Scotland, UK

Abstract. Opacity-limited fragmentation is investigated as a function of the dust-to-gas ratio and it is found that the characteristic protostellar mass M_* is metallicity-dependent. This dependence is such that, for the low metallicity gas out of which the stars of Population II formed in the halo, M_* is less than $0.1 M_\odot$.

If applicable, these theoretical considerations would predict that substellar masses have formed more frequently under the metal-poor conditions in the early Galaxy (Population II brown dwarfs). Thus the missing mass in the Galactic halo and in the dark haloes around other spirals may well reside in these metal-poor Population II brown dwarfs.

MOTIVATION

In this conference on the "Astrophysics of Brown Dwarfs" the emphasis was largely on the possible existence and discovery of a local disk population of brown dwarfs (Pop I brown dwarfs): I refer to the papers by Bahcall, Boss, D'Antona, and Hawkins in this volume. In contrast, the present note will concentrate on the possible existence and implications of a halo population of brown dwarfs (Population II brown dwarfs). This means I will discuss the effect of lowering the dust-to-gas ratio (metallicity) on the low-mass end of the Initial Mass Function (IMF). More specifically, I will discuss the characteristic stellar mass as a function of dust-to-gas ratio which I will assume scales linearly with metallicity. This characteristic mass, or mode mass as it is also called, is of order $0.1 M_\odot$ for Population I conditions in the solar neighbourhood (Scalo 1986) but may well have been different for Population II conditions in the galactic halo. If, at the low metallicities typical for halo stars, the low-mass IMF was enhanced in brown dwarfs relative to the IMF for solar composition, brown dwarfs of low metallicities (Population II brown dwarfs) may account for much if not all the dark matter hidden in the halo of our and other spiral galaxies (Faber & Gallagher 1979).

In the past, low mass objects were indeed suggested not only to solve the local but also the global "missing mass problem" in flat rotation curve spirals (Hohlfeld & Krumm 1981) as well as in galaxy clusters (Tarter & Silk 1974, Napier & Guthrie 1975). Silk (1980, p. 203) discussed the nature of dark galactic haloes in terms of low-mass stars, massive black holes, and remnants of massive stars of the first generation (Population III). Dismissing substellar objects because of the required high efficiency (> 80%) of their formation, he favoured remnants such as white dwarfs and neutron stars. However, the efficiency constraint can be overcome, i.e., the efficiency of low-mass star formation can be much lower and hence realistic (a few per cent) if one allows many generation of clouds (of order 100) forming many generation of low mass objects over the first epoch ($\sim$ 1 billion years) of galaxy formation. Then the case for substellar masses would no longer be implausible a priori.

Therefore the question becomes a question of the IMF: Is the IMF more strongly skewed towards very low mass objects during Pop II star formation, i.e., when the metallicity was low? Are substellar masses more prevalent among Population II than among Population I?

Here I draw attention to a physical argument that suggests the answer could be yes.

OPACITY-LIMITED STAR FORMATION
Several authors (Lynden-Bell 1973, Low & Lynden-Bell 1976, Rees 1976, Silk 1977a) have previously investigated the problem of the minimum protostellar mass through analytical models of opacity-limited hierarchical fragmentation. Basically two conditions must be fulfilled to determine when fragmentation must stop

(1)   the optical depth across the last fragment must exceed unity
(2)   the final fragment must satisfy the Jeans criterion (Jeans-mass)

Condition (1) ensures that the fragment starts heating up so that the Jeans mass ceases to decrease with increasing density. Conditions (1) and (2) are usually supplmented by the requirement that the compressional heating rate be equal to the cooling rate of the collapsing protostellar fragment. This requirement provides a cooling curve, i.e. the temperature as a function of density which when combined with condition (1) fixes a temperature T_f and density ρ_f separately. Then condition (2) gives a typical protostellar mass ($M_J \propto T_f^{3/2} \rho_f^{-\frac{1}{2}}$). However, there is the question whether to employ the cooling rate from an optically thin or an optically thick fragment. In principle, the optically thick case should be adopted yet the heating rate chosen is that given by a free fall collapse appropriate for the optically thin case when the cloud is still isothermal. So there are some inconsistencies in the treatment of the energy equation which reflect the circumstance that one is trying to describe the transition from one state (free-fall collapse) to another (quasi-equilibrium) without dealing with the time-dependent terms of the problem. Clearly a detailed numerical calculation is required (Boss 1984, see also Boss 1982).

It appears the optically thin case leads to the determination of the minimum Jeans mass [which is almost totally insensitive to metallicity variations (Low and Lynden-Bell 1976, Silk 1977a)] while the optically thick case leads to some maximum protostellar mass which according to

Silk (1977a) is more likely a characteristic protostellar mass or mode mass where the IMF might have its peak (cf. Zinnecker 1984). Consequently, I looked into Silk's derivation (his § III) of that characteristic mass to find a possible metallicity dependence. This produced M_* $\propto K^{1/3} T^{5/6}$ where K is the Rosseland mean grain absorption coefficient and T is the fragment gas temperature (taken to be half the central temperature) which at the high densities involved, where gas and dust are well coupled, is also the dust temperature. The central temperature is a free parameter for which Silk suggests to choose the evaporation temperature of mineral grains ($T \gtrsim 1000$ K), i.e. the maximum temperature that is still consistent with the use of grain opacity as the main source of opacity. For K appropriate to the present interstellar medium and T of order 500 K the result is $M_* \sim 0.1 M_\odot$ (see equation 20 in Silk 1977a).

Since K equals the dust-to-gas ratio times the opacity per grain (in cm^2/g) times ρ(gas) and since the dust-to-gas ratio ρ(dust)/ρ(gas) presumably varies proportional to metallicity while the opacity per grain is an intrinsic property independent of metallicity, one can write $K \propto Z/Z_\odot$ and hence obtains

$$M_* \sim 0.1 M_\odot (Z/Z_\odot)^{1/3}$$

where $Z/Z_\odot$ is the metallicity, i.e. the fractional mass in heavy elements normalised to the value in the solar vicinity ($Z_\odot \sim 0.02$).

Since the Rosseland mean absorption coefficient K enters the above expression via the radiative diffusion approximation used to estimate the cooling rate of the fragment, the result is only valid as long as there are enough dust grains to dominate the cooling rate. This imposes the limit $Z/Z_\odot \gtrsim 10^{-3}$ (approximately). For lower metallicities other sources of cooling such as line radiation from molecular hydrogen will be dominant (Silk 1977b). The important thing to note is that $Z/Z_\odot \sim 10^{-3}$ yields $M_* \sim 0.01 M_\odot$, a mode mass well below the hydrogen burning minimum mass. The implications of this result are intriguing (a systematic increase of the mass-to-light ratio of the stellar population with decreasing metallicity of the population) but were not pointed out neither by Silk (1977a) nor by Gaustad (1963) who was the first to clarify the role of dust opacity on the early stages of star formation. Gaustad concluded that the opacity he had derived was not high enough to raise the internal temperature sufficiently to establish hydrostatic equilibrium for protostars with masses greater than 0.1 $M_\odot$. Such protostars collapse in free fall all the way to nearly stellar densities while protostars with masses smaller than 0.1 $M_\odot$ will go through a slow Kelvin-Helmholtz contraction (an example for the latter is the first hydrostatic core formed during the non-homologous cloud collapse calculated by Larson 1969). The characteristic mass of order 0.1 $M_\odot$ is known as the Gaustad mass. Also, in § V of his work (page 1072), Gaustad went on to solve for the characteristic mass and already gave the result M_* $\propto K^{1/3}$ for equilibrium models to exist but did not consider the lower grain absorption coefficient that would be appropriate for low metallicity protostars (cf. his Figure 7).

OBSERVATIONAL SUPPORT

Is there any observational evidence for a Population II IMF enhanced in low-mass objects? Possibly yes. We have to distinguish between the IMF for halo field stars and the IMF in globular clusters. Mould (1982) has reviewed the Population II field star IMF and concluded that it might be steeper than the Population I field star IMF (certainly not flatter). Note that even a luminosity function that flattens toward the hydrogen burning minimum mass can lead to a rather steep mass function because of the transformation involving the derivative of the mass-luminosity relation (D'Antona & Mazzitelli 1983). Regarding globular clusters, the small observed mass-to-light ratios in the range 1-3 $M_\odot/L_\odot$ (Illingworth 1976) seem to rule out a dwarf-enhanced IMF but the observed ratios may not be the initial ones when the clusters formed, since these clusters preferentially loose low mass members during their dynamical evolution (Illingworth 1976). One of the best studied clusters is M13 which appears to be strongly dwarf-enhanced compared to a Salpeter IMF (Lupton & Gunn 1985).

A RIVAL HALO CANDIDATE POPULATION

It must be said that the brown dwarf hypothesis for dark haloes that I have tried to substantiate in this note is rivalled by another hypothesis recently suggested by Larson (1986). While he calculates that white dwarf remnants from a more intense high-mass mode of star formation in the past phases of galactic star formation could naturally explain the hidden mass in the local disk, he speculates that a vigorous burst of exclusively massive stars in a collapsing protogalaxy will leave the resulting galaxy surrounded by a dark halo consisting of the black hole remnants from those massive stars (cf. Silk 1980). It is interesting to note that both solutions to the problem of dark matter in galactic haloes are dissipative "baryonic" solutions based on star formation models (rather than dissipationless solutions based on some form of "inos"). That the dissipative solutions are so vastly different highlights the fact that there is currently no generally accepted theory of star formation yet (the picture of fragmentation limited by opacity adopted here is rivalled by Larson's picture of fragmentation limited by reaching a subsonic velocity dispersion).

OBSERVATIONAL TEST

Ultimately we will have to come up with a convincing observational test to decide between these two rival models concerning the nature of the matter in dark haloes (brown dwarfs or black holes). Such a test may involve the effects of gravitational lensing as proposed by Gott (1981) or the detection of the integrated far-infrared surface brightness of a dark halo consisting of "Jupiters" (Karimabadi & Blitz 1984). I cannot discuss "inos" here; they may well exist. But as long as we don't know for sure, we must not dismiss brown dwarfs ("substars") too readily as candidates for dark matter (cf. Gunn 1985).

ACKNOWLEDGEMENTS

I thank Drs D Emerson and J R Graham for helpful discussions. Comments from Drs J E Gaustad and J Silk were also very helpful. Thanks also to Dr A P Boss for sending his contribution as a preprint. Finally, I wish to thank the editor of these Proceedings (Dr M Kafatos) for his

support enabling me to attend the meeting and for his patience to wait
for my manuscript.

REFERENCES

Boss, A.P. (1982). A heuristic criterion for instability to fragmenta-
        tion in rotating, interstellar clouds. Astrophys. J. <u>259</u>,
        159-65.
Boss, A.P. (1984). Prototstellar formation in rotating interstellar
        clouds. IV. Nonisothermal collapse. Astrophys. J. <u>277</u>,
        768-82.
Bouchet, P., Lequeux, J., Maurice, E., Prévot, L., & Prévot-Burnichon
        (1985). The visible and infrared extinction law and the
        gas-to-dust ratio in the Small Magellanic Cloud. Astron.
        Astrophys. <u>149</u>, 330-36.
D'Antona, F. & Mazzitelli, I. (1983). Mass-luminosity relation and
        initial mass function at the faint end of the main sequence.
        Is there a real deficit of very low mass stars? Astron.
        Astrophys. <u>127</u>, 149-52.
Faber, S.M. and Gallagher, J.S. (1979). Masses and mass-to-light ratios
        of galaxies. Ann. Rev. Astron. Astrophys. <u>17</u>, 135-87.
Gaustad, J.E. (1963). The opacity of diffuse cosmic matter and the
        early stages of star formation. Astrophys. J. <u>138</u>, 1050-73.
Gott, J.R. III (1981). Are heavy haloes made of low mass stars? A
        gravitational lens test. Astrophys. J. <u>243</u>, 140-146.
Hohlfeld, R.G. & Krumm, N. (1981). An infrared search for massive
        galactic envelopes. Astrophys. J. <u>244</u>, 476-82.
Illingworth, G. (1976). The masses of globular clusters. II. Velocity
        dispersions and mass-to-light ratios. Astrophys. J. <u>204</u>,
        73-93.
Gunn, J.E. (1985). Summary talk for IAU Symposium 117. Dark matter in
        the universe. Princeton Observatory preprint 151.
Karimabadi, H. & Blitz, L. (1984). The detectability of population III
        "Jupiters". Astrophys. J. <u>283</u>, 169-73.
Larson, R.B. (1969). Numerical calculations of the dynamics of a
        collapsing protostar. Mon. Not. R. Astr. Soc. <u>145</u>, 271-95.
Larson, R.B. (1986). Bimodal star formation and remnant-dominated
        galactic models. Mon. Not. R. Astr. Soc., submitted.
Low, C. & Lynden-Bell, D. (1976). The minimum Jeans mass or when frag-
        mentation must stop. Mon. Not. R. astr. Soc. <u>176</u>, 367-90.
Lupton, R. & Gunn, J.E. (1985). M13: main-sequence photometry and the
        mass function. STScI-preprint No. 76.
Lynden-Bell, D. (1973). <u>In</u> Dynamical Structure and Evolution of Stellar
        Systems, eds. G. Contopoulos, M. Henon, & D. Lynden-Bell,
        (Geneva: Geneva Observatory), p.131.
Mould, J. (1982). Stellar populations in the Galaxy. Ann. Rev. Astr.
        Astrophys. <u>20</u>, 91-115.
Napier, W. McD. & Guthrie, B.N.G. (1975). Black dwarf stars as missing
        mass in clusters of galaxies. Mon. Not. R. Astr. Soc. <u>170</u>,
        7-14.
Rees, M.J. (1976). Opacity-limited fragmentation and the masses of
        protostars. Mon. Not. R. Astr. Soc. <u>176</u>, 483-486.
Scalo, J.M. (1986). The initial mass function. Fund. Cosm. Phys. (in
        press).

Silk, J. (1977a). On the fragmentation of cosmic gas clouds. II.
         Opacity-limited star formation. Astrophys. J. 214, 152-60.
Silk, J. (1977b). On the fragmentation of cosmic gas clouds. I. The
         formation of galaxies and the first generation of stars.
         Astrophys. J. 211, 638-48.
Silk, J. (1980). In Star Formation, eds. A. Maeder and L. Martinet
         (Saas-Fee: Geneva Observatory), Chap. VIII.
Tarter, J. & Silk, J. (1974). Current constraints on hidden mass in
         the Coma Cluster. Quart. J. Roy. Astr. Soc. 15, 122-40.
Zinnecker, H. (1984). Star formation from hierarchical cloud frag-
         mentation: a statistical theory for the log-normal initial
         mass function. Mon. Not. R. Astr. Soc. 210, 43-56.

DISCUSSION

J. TARTER: The problem with keeping a spherical population can be
avoided by making them early as Population III! But until $Z/Z_0 \gtrsim 10^{-3}$
the minimum mass is 0.3 → 3 M_0 (from the 1977a paper by Silk) so you
are stuck with a dissipating flattened system for the low mass fragments.
H. ZINNECKER: Not quite. You can make a small amount of metals (Z/Z_0
$\gtrsim 10^{-3}$) in Population III stars before galaxy formation so that the
protogalactic cloud does not start out to collapse with zero metallicity.
Then enough grains exist that early star formation in the halo is rather
similar to star formation later in the disk, except that the lower
dust-to-gas ratio of the halo material drives fragmentation to lower
masses, as argued above.
B. TARTER: How confident are you of the opacity extrapolations as a
function of metallicity?
H. ZINNECKER: Fairly confident. Comparing observations of the dust-to-
gas ratio in the Galaxy to that in the Magellanic Clouds confirms
(Bouchet et al. 1985) that this ratio scales linearly with metallicity,
as assumed here. Moreover, even if grain sizes were metallicity-dependent, the opacity per grain would be metallicity-independent since the
opacity per grain is approximately independent of grain size at the
relevant long infrared wavelengths [cf. equation (16) in Silk (1977a)].

HIGH MASS PLANETS AND LOW MASS STARS*

D.J. Stevenson
California Institute of Technology, 170-25, Pasadena,
California 91125 USA

ABSTRACT. There are three main ingredients in brown
dwarf theoretical models: (i) The equation of state;
(ii) The entropy equation, which relates the internal and
atmospheric thermodynamic states; and (iii) The
atmospheric boundary condition (the infrared opacity).
It is argued that the first two ingredients are very well
understood. The opacity is less well understood and the
major unresolved problem. Simple scaling laws are
described and discussed for the relationships between
luminosity (L), mass, opacity and age (t), assuming no
thermonuclear energy sources. In the limit of extreme
degeneracy, $L \propto t^{-1.25}$. However, detectable brown dwarfs
(including VB8B) are still significantly contracting
(i.e. actual radius ~10% larger than the zero temperature
limit). As a consequence, dlnL/dlnt ~ −1.0 to −1.1 at
this epoch for VB8B. Large opacity increases the effect
of non-degeneracy. Complicating factors in brown dwarf
evolution (super- or sub-adiabaticity, Debye cooling,
freezing, differentiation, variable opacity) are discussed but only the latter two seem like to be important.

INTRODUCTION

This contribution deals with bodies which never had significant thermonuclear energy sources. Since deuterium burning is only a minor
delaying tactic on the path toward degenerate cooling if the mass is
less than 0.08 $M_\odot$, we consider all masses up to this limit. These
''brown dwarfs'' are like Jupiter only simpler. We understand Jupiter
well (c.f. Stevenson, 1982) so we should understand brown dwarfs even
better. A contrary impression has arisen, primarily because of the

_____

*Contribution number 4288 from the Division of Geological and Plane-
tary Sciences, California Institute of Technology, Pasadena, CA 91125.

large discrepancies between the results quoted by Tarter (1975) and the derived scaling laws of Stevenson (1978). In fact, these discrepancies arose not because of inadequacies in the basic physics, but because Tarter constructed purely empirical scaling relations (to satisfy the numerical results of Graboske and co-workers) which did not describe fully degenerate configurations, whereas Stevenson developed an analytical theory which is designed to work best for the degenerate phase.

In this contribution, the ingredients of the theory are described and discussed, so that the reader can judge for him or herself where the uncertainties lie. An analytical model, based on a polytrope (n = 3/2) is developed which encompasses both degenerate and non-degenerate regimes and forms the basis of improved (but more complicated) scaling laws. More complete modeling efforts are also discussed briefly.

The beginning assumptions are:

(i)   Brown dwarfs are homogeneous and close to cosmic composition.

(ii)   These bodies are fully convective, except for the outermost radiative layer.

(iii)   Convective efficiency is high so that the entire body is very close to isentropic.

(iv)   The atmospheric radiative boundary condition can be characterized by a single (broad band) opacity.

In fact, none of these assumptions is likely to be strictly correct and each will be challenged as this paper develops. Nevertheless, the uncertainties turn out to be small, except in the opacity which remains a problem, especially for relating what the theorist calculates to what the observer measures.

EQUATION OF STATE

One of the reasons why brown dwarfs are simpler than giant planets is the fact that their equation of state is within 10% of $p \propto \rho^{5/3}$, almost everywhere and irrespective of temperature (i.e. evolutionary age). The total pressure can be approximated as a sum of Fermi, exchange, Coulomb, thermal electronic, and thermal ionic terms:

$$p = p_{Fermi} + p_{exch} + p_{coul} + p_{th,el} + p_{th,ion} \qquad (1)$$

$$P_{Fermi} = 9.92 \times 10^{12} \, (\rho/\mu_e)^{5/3} \; dyne.cm^{-2} \tag{2}$$

$$P_{exch} = -2.06 \times 10^{12} \, (\rho/\mu_e)^{4/3} \, f_{ex} \; dyne.cm^{-2} \tag{3}$$

$$P_{coul} = -4.04 \times 10^{12} \, (\rho/\mu_e)^{4/3} \, \bar{z}^{5/3} \, f_c \; dyne.cm^{-2} \tag{4}$$

$$P_{th,el} = f_e n_{el} k_B T = f_{el} \, (\rho/\mu_e m_p) \, k_B T \tag{5}$$

$$P_{th,ion} = f_i n_i k_B T = f_i (\, \rho/\bar{A} m_p \,) \, k_B T \tag{6}$$

$$f_{ex} \cong \min (1, \, 2.4/\psi)$$

$$f_{el} \cong \min (1, \psi /2.4) \tag{7}$$

$$f_c \cong \min (1, \, \Gamma^{1/2}) \tag{8}$$

$$f_i \cong 1 + (\Gamma/300) \tag{9}$$

$$\psi = 1.594 \times 10^{-5} \, T/(\rho/\mu_e)^{2/3} \tag{10}$$

$$\Gamma = 2.27 \times 10^5 \, (\rho/\mu_e)^{1/3} \, \bar{z}^{5/3}/T \tag{11}$$

where ρ is the density in g.cm^{-3}, μ_e is the mean molecular weight per electron ($\mu_e \sim 1.15$ for cosmic composition), $\bar{z}$ is the mean nuclear charge (~ 1.08), k_B is Boltzmann's constant, T is the temperature in Kelvin, m_p is the mass of the proton, $\bar{A}$ is the mean atomic mass, ψ is a degeneracy parameter ($\psi \ll 1$ implies degeneracy), and Γ is a plasma parameter ($\Gamma \ll 1$ implies the Debye–Huckel regime, $\Gamma \gg 1$ implies a strongly coupled Coulomb plasma, $\Gamma \gtrsim 180$ implies a solid lattice). The fudge parameters f_e, f_c, and f_i are given approximate but adequate forms here; for a lengthy discussion, see, for example, DeWitt (1969).

In an isentropic, nearly ideal electron gas, $\gamma = (d\ln T/d\ln \rho) \simeq 2/3$ and all the terms in equation (1) except p_{exch} and p_{coul} scale approximately as $\rho^{5/3}$. As we discuss below, even dense Coulomb plasmas have $\gamma \sim 0.6$ and this scaling still applies. Under the conditions of interest ($\rho \sim 10^2 - 10^3$ g.cm^{-3}), $p_{exch} + p_{coul}$ is over a factor of ten smaller than the sum of all the other terms in equation (1), irrespective of T.

Equation (1) breaks down in a thin layer which underlies the radiative layer of the atmosphere and overlies the fully ionized interior. In this region, molecules and atoms undergo dissociation and ionization. This layer is typically less than 10^{-3} of the total mass and large variations in its treatment have negligible effect on the resulting static and evolutionary properties of the brown dwarf.

THE ENTROPY

Except in the ideal gas limit, the internal entropy of a brown dwarf is not analytically calculable and one must rely on Monte Carlo simulations (e.g. Hubbard and DeWitt, 1976) or variational models of the fluid state (Stevenson, 1975). In general, the internal entropy can be expressed in the form

$$S_{int} = A \ln (T/\rho^{\gamma}) + B \qquad (12)$$

where A, B, and γ are approximately constant. If the entropy is in units of Boltzmann's constant per nucleus, then $A \simeq 2.2$, $\gamma \simeq 0.63$, and $B \simeq -11.6$ (chosen to fit Stevenson, 1975, for a cosmic H/He). This formula includes the electronic contribution, but fails at low temperatures. However, this failure occurs only if hydrogen freezes, a situation never encountered in models of interest. The uncertainties in S_{int} translate to $\lesssim 30\%$ uncertainty in temperature.

For $T \lesssim 2000$ K, the atmospheric entropy is dominated by the translation and rotation of hydrogen molecules:

$$S_{atm} \simeq 1.27 \ln (T/\rho^{0.42}) - 3.0 \qquad (13)$$

for cosmic H/He. Above T ~ 2000 K, dissociation of hydrogen molecules becomes increasingly important.

THE OPACITY

If a single, broad-band opacity, κ, characterizes the IR emission of the atmosphere, then

$$p_e \simeq g/\kappa \qquad\qquad\qquad (14)$$

where p_e is the pressure at optical depth unity and g is the local gravitational acceleration. A numerical factor somewhat different from unity could be included on the R.H.S. but in practice, κ is too uncertain to justify bothering about this factor. For the same reason, we will assume that the atmospheric entropy (equation 13) can be evaluated using the thermodynamic state at optical depth unity, even though this level is in a mildly stable part of the atmosphere. (For Jupiter, this assumption is equivalent to an error of only a few percent in κ.)

Figure 1 illustrates the problem of deciding what value of to use. If grain opacity were ignored, $\kappa \sim 10^{-2}$–10^{-1} $cm^2.g^{-1}$ but then a grey atmosphere assumption may be invalid (Lunine et al., this conference). In fact, as they argue, grain opacity seems to be required by the data for VB8B. The problem with grain opacity is that its magnitude is difficult to estimate because of large uncertainties in the grain size. It is certainly not correct to use tables such as those provided by Alexander (1975) for 0.1 μ particles, since the grains or droplets in brown dwarf atmospheres are expected to grow much larger, on average. However, one does not know the size spectrum, so the only safe statement is that the opacity is bounded above by calculations based on very small grain size. Furthermore, the grain opacity will diminish at low temperature (rather than be ''frozen in'', as Alexander assumed) because a cloud deck forms and the radiating level will eventually be above the clouds. Fortunately, as the simple model below illustrates, the opacity enters rather weakly in the determination of brown dwarf evolution, unless it changes very rapidly with temperature.

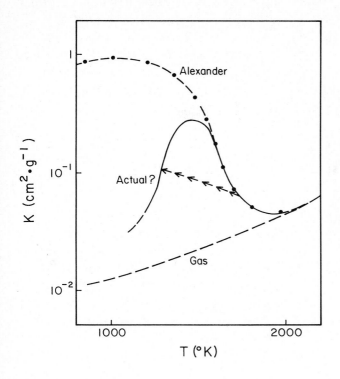

Figure 1. Approximate
opacity curves (actually
dependent on pressure as
well as temperature) for
an atmosphere that has no
grains (gas only), has
0.1 μ grains ''frozen
in'' (Alexander, 1975) or
allows for large grains
and development of a
cloud deck (curve labeled
realistic).

ANALYTICAL MODEL

The analytical model neglects thermonuclear reactions and is constructed from three ingredients: (i) the relationship between radius,
mass, and entropy for an adiabatic n = 3/2 polytrope; (ii) the
entropy equation linking the effective temperature to the central
temperature; (iii) the first law of thermodynamics, linking the
contraction to the luminosity.

Consider, first, the radius–mass–entropy relationship. If
we neglect exchange and Coulomb pressure contributions then

$$p = K\rho^{5/3}$$

$$K = K_o [1 + aT/\rho^{2/3}] \qquad \text{(non–degenerate)}$$

$$K = K_o [1 + aT/2\rho^{2/3} + b(T/\rho^{2/3})^2] \qquad \text{(degenerate)} \quad (15)$$

where a and b are constants, K_0 is the zero temperature limit of K, and $T/\rho^{2/3}$ is assumed constant, in accordance with the discussion above. Although the corrections due to non-degeneracy are important, it turns out that the ''degenerate'' limit of equation (15) is satis-factory for $R \lesssim 3R_0$, where R is the actual radius and R_0 is the zero temperature radius for the same mass. It then follows from the solution for an n = 3/2 polytrope (Clayton, 1968) that

$$R = R_0 [1 + \frac{a}{2} (T/\rho^{2/3}) + b (T/\rho^{2/3})^2] \qquad (16)$$

$$R_0 = 2.2 \times 10^9 (M_\odot/M)^{1/3} \qquad (17)$$

Since the <u>shape</u> of the density profile is invariant, this can be rewritten in the form

$$\frac{R}{R_0} = 1 + \frac{T_c}{T_0} (\frac{R}{R_0})^2 + 0.4 (\frac{T_c}{T_0})^2 (\frac{R}{R_0})^4 \qquad (18)$$

$$T_0 \cong (4.9 \times 10^8 \text{ K}) (\frac{M}{M_\odot})^{4/3} \qquad (19)$$

where T_c is the central temperature. Equation (18) predicts that as the star contracts (R/R_0 decreasing), the temperature first rises to reach a maximum ~0.18 T_0 and then decreases. The requirement that the maximum temperature exceed ~3 x 10^6 K for hydrogen burning implies a minimum main sequence mass of ~0.08 $M_\odot$. For our purposes, it is sufficient to approximate equation (18) by $T_c/T_0 = R_0/R - (R_0/R)^2$.

Turning now to the implications of isentropy, equations (12), (13), and (14) can be combined, together with $\rho_c = \rho_0(R_0/R)^3$, $g = (GM/R_0^2)(R_0/R)^2$, and $p_e = \rho_e kT_e/\mu$ ($\mu \cong 3.75 \times 10^{-24}$ g) to give

$$T_e = 3.8 \times 10^{-6} T_c^{1.22} (M/M_\odot)^{-1.05} (R/R_0)^{1.7} (K/10^{-2})^{-0.29} \qquad (20)$$

Notice that a given fractional change in T_e must be matched by a comparable fractional change in T_c, if   is constant. This is important for a qualitative understanding of the luminosity-time relation derived below.

From the first law of thermodynamics, and the assumption of hydrostatic equilibrium, the luminosity L is given by

$$L = -\frac{d}{dt} \int_o^M (E + \frac{p}{\rho^2} \frac{d\rho}{dt})\, dm \qquad (21)$$

where E is the internal energy and the second term represents the rate of release of gravitational energy. If we subdivide the pressure and internal energy into zero temperature components plus thermal corrections then

$$L = -\frac{d}{dt} [\int_o^M (C_v T + \frac{p_{th}}{\rho^2} \frac{d\rho}{dt})\, dm] \qquad (22)$$

where C_v is the specific heat at constant volume and p_{th} is the thermal contribution to the pressure. For $C_v \simeq 2k_B/\mu$ ($\mu \simeq 1.08\ m_p$) and $p_{th} \simeq \rho k_B T/\mu$, we find

$$L \simeq -\frac{0.6\, M\, k_B\, T_o\, R_o}{R^3 \mu} (R + R_o) \frac{dR}{dt} \qquad (23)$$

where the coefficient (0.6) is specific to the n = 3/2 polytrope. Since $L = 4\pi R^2 \sigma T_e^4$, equations (18), (20), and (23) lead to a differential equation of the form

$$x^2(x - 1)^{4.88} = -\tau(1 + x)\, dx/dt \qquad (24)$$

where $x = R/R_o$ and τ is a time constant (dependent on M and K). The asymptotic solution (x -> 1) of this equation is

$$x = 1 + \varepsilon - 0.52 \, \varepsilon^2 \ldots.$$

$$\varepsilon \equiv \left(\frac{3.88 \, t}{\tau} \right)^{-0.258}$$

$$\simeq 0.06 \left(\frac{5 \times 10^9 \, yr}{t} \right)^{0.258} \left(\frac{M}{0.08 \, M_\odot} \right)^{0.17} \left(\frac{K}{10^{-2}} \right)^{0.3} \quad (25)$$

from which the following scaling laws immediately follow:

$$\frac{L}{L_\odot} \simeq 1.5 \times 10^{-5} \left(\frac{5 \times 10^9 \, yr}{t} \right)^{5/4} \left(\frac{M}{0.08 \, M_\odot} \right)^{5/2} \left(\frac{K}{10^{-2}} \right)^{0.3} \left(1 - \frac{7}{2} \varepsilon \right)$$

$$T_e \simeq 1420 \left(\frac{5 \times 10^9}{t} \right)^{5/16} \left(\frac{M}{0.08 \, M_\odot} \right)^{0.79} \left(\frac{K}{10^{-2}} \right)^{0.075} \left(1 - \frac{11}{8} \varepsilon \right) \quad (26)$$

where some irrational exponents have been approximated by nearby fractions.

Several interesting features emerge from this scaling law. Consider, first, the asymptotic, fully degenerate limit (ε negligible). The luminosity ($\sim T_e^4$) is then balanced by the decrease of internal thermal energy ($\sim dT_c/dt$). Using equation (20) ($T_e \sim T_c^{1.22}$), it follows immediately that $L \sim t^{-1.25}$. As required by the finiteness of the energy supply, $\int^\infty L dt$ is finite. Tarter's scaling law ($L \sim t^{-0.84}$) is invalid because it violates this fundamental constraint. (This difficulty cannot be avoided by appealing to Debye cooling or freezing because they do not happen for any effective temperature that can be attained in the age of the Universe.) However, corrections due to finite temperature (finite ε) are non-negligible in general. From equation (26), $-dlnL/dlnt \simeq (5/4)-(7/4)\varepsilon$, indicating that even at $\varepsilon \sim 0.1$, there is a substantial deviation from the asymptotic behavior. In fact, detectable bodies (including VB8B) are likely to have $\varepsilon \sim 0.1$ or larger; this casts doubt on the usefulness of the scaling laws. Another way to appreciate this difficulty is to demand that T_e decrease monoptically with time; accordingly the above approximation must fail for $\varepsilon \gtrsim 0.26$. Other factors being equal, the biggest influence on ε is the opacity; large opacity implies that the approach to degeneracy is greatly delayed. Notice that the formula for L involves K directly and also indirectly (through ε); these two dependences tend to counterbalance, as expected since the time-integral of the luminosity should be independent of K and determined only by the Virial theorem and the first law of thermodynamics.

Application to VB8B yields satisfactory agreement for a wide range of masses, provided no constraint is imposed on the evolution time. If we require 1×10^9 yr $\lesssim t \lesssim 5 \times 10^9$ yr then the mass of VB8B is in the range $0.04 \lesssim M/M_0 \lesssim 0.08$, with the middle of this range being most plausible. Figure 2 shows theoretical evolution curves and a comparison with VB8B.

COMPLICATIONS

(a) Deviations from an n = 3/2 polytrope: This is increasingly important as the mass becomes lower, but the main effect

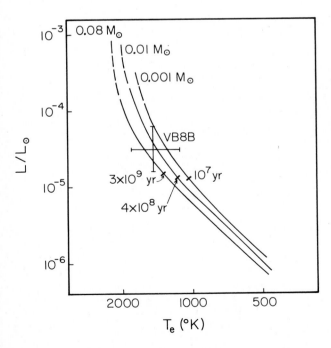

Figure 2. Theoretical Hertzsprung–Russell diagram for three masses, with the VB8B data point superimposed. Times indicate elapsed age for that mass and luminosity.

is on the radius—mass relationship at T = 0; the general properties
of the scaling laws above are preserved. Finite temperature corrections become <u>less</u> important at lower mass (even at early times)
because if P $K\rho^{1+(1/n)}$ then R ~ $K^{n/(3-n)}$ which is a weaker dependence
of R on T when n < 3/2. At the mass of Jupiter (0.001 $M_{\odot}$), the best
approximation to the polytropic index is n $\approx$ 1.0.

(b) <u>Superadiabatic</u> <u>Convection</u>: According to mixing
length theory, the heat flux transported by convection is about

$$F \overset{\sim}{} 0.1 \ (\alpha T)\rho \ C_p \ v_s \ T \ \eta^{3/2} \tag{27}$$

where α is the coefficient of thermal expansion, v_s is the local sound
speed, η is the fractional superadiabaticity and it is assumed that
mixing length ~ pressure scaleheight. For an object like VB8B, is
largest just below optical depth unity and $\eta \lesssim 0.03$. This is a
negligible effect.

(c) <u>Subadiabatic</u>, <u>Conductive</u> <u>Core</u>: In the deep interior,
degenerate electrons provide a high thermal conductivity k $\overset{\sim}{} 10^8$
$\rho^{2/3}$ cgs (Stevenson and Ashcroft, 1974). As a result, conduction
becomes important, especially in the more massive bodies (Stevenson,
1978). However, this has only a small effect on the luminosity-time
relation, at least for t $\lesssim 5 \times 10^9$ yr.

(d) <u>Debye</u> <u>Cooling</u>: If T $\lesssim \theta_D$, the Debye temperature,
in the deep interior then the specific heat drops dramatically (C_v ~
$(T/\theta_D)^3$, T $\lesssim \theta_D$) and the luminosity would be greatly affected.
This phenomenon has been extensively studied for white dwarfs (see
discussion in Shapiro and Teukolsky, 1983, Ch. 4). However, $\theta_D \propto$
$\rho^{1/2}$ whereas $T_c/\rho_c^{2/3}$ is comparable for brown dwarfs of comparable
age. As a result, Debye cooling should be most important at <u>low</u>
masses. Since it has not yet happened in Jupiter (Stevenson, 1982),
it clearly has not happened in more massive bodies. In fact,

$$\frac{T_c}{\theta_D} \overset{\sim}{} 20 \ (\frac{M}{0.08 \ M_{\odot}})^{0.3} \tag{28}$$

at t ~ 5×10^9 yr, K ~ 10^{-2} cm^2/g.

(e) _Freezing_: This is even less likely because the melting temperature $T_M \propto \rho^{1/3}$ for a Coulomb plasma. In fact, $T_M/\theta_D \simeq \rho^{1/6}$ (so $T_M \sim \theta_D$ at $\rho \sim 1$, and $T_M < \theta_D$ at higher densities).

(f) _Differentiation_: The pressure at the center of a 0.06 M_0 brown dwarf is comparable to the pressure at the center of the Sun; the temperature is about one order of magnitude lower. 'Bare' ions of charge $Z \gg 1$ have a solubility in metallic hydrogen of order $\exp(-T_z/T)$, where

$$T_z \stackrel{\sim}{} 10^4 \; Z^{7/3} \tag{29}$$

(Stevenson, 1976). For example, the critical temperature for the unmixing of Fe^{24+} is $\sim 6 \times 10^6$ K. For an ion with atomic abundance $\sim 10^{-4}$, insolubility and core formation occurs in a brown dwarf ($T_c \sim 1 \times 10^6$) provided $Z \gtrsim 20$. This is probably achieved for iron, although the situation is complicated by the presence of at least a few bound states. Actually, bound states can _increase_ the likelihood of insolubility in some circumstances (Stevenson, in preparation). Although brown dwarfs may form iron cores (and the iron may possibly extract some other high Z elements also), this is unlikely to affect brown dwarf evolution, except indirectly by reducing the abundance of elements needed to provide grain opacity in the atmosphere. Unless the body has a greatly enhanced abundance of high Z elements relative to cosmic abundance, the resulting core will be small and the gravitational energy of its formation will be negligible (unlike the situation in Jupiter or Saturn, see Salpeter, 1973).

(g) _Variation of Opacity_: The scaling relations are only applicable for constant K. If the dependence of K on temperature is strong enough then very different evolutions can occur. Consider equation (20), $T_e \sim T_c^{1.22} K^{-0.29}$. If $d\ln K/d\ln T_e < -3.5$, then $d\ln T_e/d\ln T_c < 0$, an impossible situation to sustain (central temperature goes up as the effective temperature drops). This behavior was previously noted in a somewhat different context by Rappaport et al. (1982). The star responds by dropping to a much lower T_e on the other side of the opacity peak (Fig. 1). Since the energy stored within the star is unaltered, it can then radiate for a longer period of time at this reduced T_e. This may be relevant to the interpretation of VB8B.

DETAILED MODELS

Aside from the early and somewhat incomplete modeling of Grossman and
Graboske (1973; see also Tarter, 1975), the only detailed numerical
models of brown dwarfs are those of D'Antona and Mazzitelli (1985),
Nelson et al. (1985), and the work described at this conference.
Nelson and co-workers were the first to make a careful comparison
between the theory and the observational characteristics of VB8B.
Although there are some significant differences among these models
(mainly in the entropy equation and opacity) the essential features of
the scaling laws described above are confirmed. A more careful comparison between simple analytical models and the detailed numerical
models of Hofmeister and Stevenson (in preparation) reveals that the
fully degenerate limit is rather slowly attained. For this reason,
analyses of the luminosity function and detectability of brown dwarfs
(c.f. Probst, 1983) may require more accurate (non-power law) descrip-
tions of the luminosity-time relationship. The calculations of
Hofmeister and Stevenson also indicate the possibility of a ''jump''
in T_e and L because of the strong inverse dependence of opacity on
temperature at $T \lesssim 2000$ K.

CONCLUDING COMMENTS

Basically, brown dwarf theory is in good shape. There are admittedly
quite large uncertainties in the opacity and this translates into
significant uncertainties in the comparison of observables and
theoretical predictions, especially if brown dwarfs are not grey
bodies. These uncertainties in opacity also imply substantial
uncertainties in the luminosity (but not the mass) at the bottom end
of the main sequence (D'Antona and Mazzitelli, 1985). However, the
interiors of brown dwarfs offer no major theoretical challenges
except, possibly, the question of differentiation. The identification
of VB8B as a brown dwarf of mass $0.06(\pm0.02)M_\odot$ seems reasonable
although a residual doubt persists because of the absence of adequate
data to characterize how ''non-grey'' the atmosphere is.

The biggest challenges for the future lie not in theory
but in observation: we need more than two colors. We need spectra.
We need more bodies. Refinement of the theory does not seem to be a
compelling task until this happens.

REFERENCES

Alexander, D.R. (1975) Low temperature Rosseland opacity tables. Astrophys. J. Suppl. **29**, 363-374.

Clayton, D.D. (1968) Principles of Stellar Evolution and Nucleosynthesis, McGraw-Hill, p. 160.

D'Antona, F. and Mazzitelli, I. (1985) Evolution of the very low mass stars and brown dwarfs. I. The minimum main sequence mass and luminosity. Astrophys. J. **296**, 502-513.

DeWitt, H.E. (1969) Statistical mechanics of dense ionized gases. In Low Luminosity Stars, ed. S.S. Kumar, publ. Gordon and Breach (NY), pp. 211-222.

DeWitt, H.E. and Hubbard, W.B. (1976) Statistical mechanics of light elements at high pressure. IV. A model free energy for the metallic phase. Astrophys. J. **205**, 295-301.

Grossman, A.S. and Graboske, Jr., H.C. (1973) Evolution of low mass stars. V. Minimum mass for the deuterium main sequence. Astrophys. J. **180**, 195-198.

McCarthy, D.W., Probst, R.G., and Low, F.J. (1985) Infrared detection of a close cool companion to Van Biesbroeck 8. Astrophys. J. **290**, L9-13.

Nelson, L.A., Rappaport, S.A., and Joss, P.C. (1985) On the nature of the companion to Von Biesbroeck 8. Nature **316**, 42-44.

Probst, R.G. (1983) Looking for brown dwarfs. IAU Colloq. 76. ''The Nearby Stars and the Stellar Luminosity Function.''

Rappaport, S., Joss, P.C., and Webbink, R.F. (1982) The evolution of highly compact binary stellar systems. Astrophys. J. **254**, 616-640.

Salpeter, E.E. (1973) On convection and gravitational layering in Jupiter and in stars of low mass. Astrophys. J. **181**, L83-86.

Shapiro, S.L. and Teukolsky, S.A. (1983) Black Holes, White Dwarfs and Neutron Stars, publ. John Wiley and Sons, N.Y., 645 pp.

Stevenson, D.J. (1975) Thermodynamics and phase separation of dense, fully-ionized hydrogen-helium fluid mixtures. Phys. Rev. **12B**, 3999-4007.

Stevenson, D.J. (1976) Miscibility gaps in fully pressure-ionized binary alloys. Phys. Lett. **58A**, 282-284.

Stevenson, D.J. (1978) Brown and black dwarfs: Their structure, evolution and contribution to the missing mass. Proc. Astron. Soc. Australia **3**, no. 3, 227-229.

Stevenson, D.J. (1982) Interiors of the giant planets. Ann. Rev. Earth Planet. Sci. **10**, 257-295.

Stevenson, D.J. and Ashcroft, N.W. (1974) Conduction in fully ionized liquid metals. Phys. Rev. **9A**, 782-789.

Tarter, J.C. (1975) Brown dwarf stars and how they grow old. In unpublished Ph.D. thesis, U.C. Berkeley.

DISCUSSION

HUBBARD: The predicted unmixing temperature which you present for Fe^{24+} is ~6 x 10^6 K. This is very much higher than the calculated central temperatures in these bodies. Is your assumed ionization state consistent with the actual temperatures? What is the ionization state of the pure (or almost pure) Fe phase?

STEVENSON: At a hydrogen density of 10^3 g.cm^{-3}, a dilute solution of iron in hydrogen will pressure-ionize to Z ~ 20-24, based on pseudopotential calculations. Pure iron may be slightly less highly ionized (although its density is ~1 x 10^4 g.cm^{-3}, close to the pressure ionization estimate in Clayton [1968], p. 154). In general, my calculations indicate greater unmixing as the temperature is decreased, even when some of the ionization is temperature induced. The models of Stevenson [1976] should not be used when bound states are present; more recent work (in progress) suggest that bound states can sometimes enhance the insolubility.

BROWN DWARFS: CONFERENCE SUMMARY

John N. Bahcall
Institute for Advanced Study, Princeton, NJ 08540

Some of the major observational and theoretical issues in the study of brown dwarfs are reviewed. It is concluded that all of the unseen local disk matter could be in the form of brown dwarfs without conflicting with any available observations.

INTRODUCTORY REMARKS

There has been a lot of discussion at this conference over the precise definition of brown dwarfs and whether a particular object does or does not qualify. I think that this is not a very important issue. The physics of brown dwarfs is very similar, for any plausible definition, to the physics of late M-dwarfs. Both must exist; both are fascinating objects to study. There is no particular reason for the *mass-function* to be discontinuous at the boundary of hydrogen burning. There must be lots of both M-dwarfs and brown dwarfs out there. Our task is to find them.

Let me give you my personal definition of a brown dwarf in terms of the *average* rate (over a Hubble time) at which a bound stellar-like mass generates nuclear energy, $L_{nuclear}$, and gravitational energy, $L_{gravity}$. I call an object a brown dwarf if it satisfies:

$$L_{nuclear} \ll L_{gravity}. \tag{1}$$

This is a relatively lax definition since it allows for some nuclear energy generation. Also, since equation (1) refers to the average rate of energy generation , the definition includes objects in which deuterium burning is temporarily the dominant mode of energy generation.

Unfortunately, equation (1) cannot be applied directly in practical situations. The reason why is that we can measure only the *total* luminosity, not the separate contributions of gravitational and nuclear energy generation. However, we are saved by the fact that theoretical models divide sharply in stellar mass according to whether equation (1) is satisfied or its converse. There is a general consensus among theorists at this conference that, for Population I stars, equation (1) is equivalent to

$$M \leq 0.08 M_\odot. \tag{2}$$

It would be nice to know quantitatively how the number in equation (2) is affected both by rotation of the initial mass and by possible accretion from the surrounding medium.

At present, equation (2) is the *only* way we can establish definitely that an object is a brown dwarf. In the future, we may have enough experience with brown dwarfs that we can use secondary criteria, but for now we have to make do with equation (2).

There are a number of different ways to find brown dwarfs. We have heard wonderful review talks in the conference describing how people are using the following techniques: classical astrometry, infrared speckle interferometry, kinematic surveys, panoramic surveys, imaging in the infrared, measuring accurate radial velocities, and surveys in space (with IRAS, HST, or SIRTF). These talks should be studied carefully and savored: they represent a very high level of technical and astronomical skill.

I will not make any references by name in this talk since all of the topics I will discuss will be represented in the papers that will appear with proper references in the book version of this conference. I will instead give you my impressions of some of the major scientific issues raised at this conference. Before doing that, however, I want to comment on the known number of brown dwarfs and the reasons why we are studying them.

By anybody's definition, the ratio of the number of papers in this field, N_{papers}, to the number of brown dwarfs that have been identified so far with certainty, $N_{identified}$, satisfies:

$$N_{papers}/N_{identified} \gtrsim 10^{+2}, \tag{3}$$

The number of research workers also far exceeds the number of certain identifications. As I count them, there are between 10 and 100 of us for every certain brown dwarf described in the Ap. J.

Why are we all so interested in these objects? I know of three reasons. 1). They *may* account for the missing mass in the disk. 2). Their physics is fascinating and, to a satisfying extent, calculable. 3). *No one else has found very many of them.* I have a feeling that the third reason adds considerable zest to the observational searches.

SOME IMPRESSIONS

The observational characteristics, other than mass, of brown dwarfs are very uncertain at present. They can be mixed together with normal stars in certain domains of color or even total luminosity. The situation does not seem likely to get much better very quickly, since the way a brown dwarf of a given mass (or luminosity) will appear to an observer depends upon its atmospheric opacity. There are large uncertainties here, including the crucial effects of huge grains, of pressure broadened gas opacities, and of unknown surface compositions. We desperately need measurable atmospheric criteria which can be related to the theoretical quantities of mass and luminosity. Incidentally, I was shocked when I realized that the determination of the effective temperature and luminosity of everyone's favorite brown dwarf, VB8 B, depends upon only two broad-band measurements of an atmospheric spectrum that may be shot full of (unknown) molecular features. We urgently need multi-band photometry for VB8 B and other prime brown dwarf candidates. Ideally, one could make nearly bolometric measurements of a significant sample, say ten, of the faintest M dwarfs and the best brown dwarf candidates in order to bypass the uncertainties that arise from using colors to determine effective temperatures.

The most certain identifications of brown dwarfs are likely to come, in the forseeable future, from classical *astrometric* measurements coupled with the infrared imaging of faint companions (perhaps by speckle interferometry). We need more such measurements coupled with detailed estimates of the allowed range of masses that are permitted by the observations. We should continue also to search for brown dwarfs with any observational technique that seems feasible. Perhaps, they will turn up in droves as the result of very accurate radial velocity measurements or some other technique that is not currently being employed. Incidentally, I was charmed by the simplicity and power that results from putting an HF absorption tube in front of a spectrograph. This system elegantly eliminates many of the systematic uncertainties because the calibration lines suffer the same effects as the stellar spectrum.

Large surveys will be required to answer statistical questions like: How much mass is there locally in the form of brown dwarfs? Kinematic surveys have yielded many of the most interesting candidates so far. However, such surveys are necessarily biased because they discriminate against low-velocity populations. Suppose brown dwarfs cool rapidly. Then all of the visible brown dwarfs are young and move relatively slowly. If this is the case, we are going to be in a lot of trouble with kinematically selected samples.

Panoramic surveys are in principle easiest to analyze. But, the results are not usually presented in a form in which we can appreciate readily the uncertainties. Suppose that a survey is done in two colors, for example, R and I. Then what is measured is the number of objects per square degree in a given color range R-I for different limiting R and I magnitudes. A few different observers ought to measure the same quantities and compare their results in the same *observed* plane, say measured number of stars versus R-I. Instead, as we have seen at this conference, all too often each observing team has a different set of colors and they compare their results in the *theoretical* plane of luminosity function versus absolute magnitude. It is very hard to evaluate the experimental uncertainties in a theoretical plane.

What does one have to do to order to go from a theoretical mass density, say 0.1 $M_\odot$ per cubic parsec, to a predicted number of stars per square degree on the sky? We need to know, first, a mass function as it depends upon galactic age, $N(m,t)$. Theory really gives us no reliable handle on this problem at present. Most of the stars could be born with typical masses of 0.02 $M_\odot$, in which case we are only observing the far tail of the distribution. *In order to proceed, we need to have faith that the missing mass will be in the observable region.* If we assume that we know the shape of the mass function, and it is favorable, we need next to relate mass to total luminosity and galactic age. Fortunately, this step can be done relatively reliably with existing theoretical models. But, all hell breaks loose at the next stage. What are the broad band colors of an object with a specified total luminosity and mass? Theorists are not willing to stick their necks out on this question since it depends upon unknown opacities, the nature and amount of huge grains, and the undetermined surface composition. Observers quite correctly say they cannot be asked to establish the luminosity-color characteristics of a population that has not yet been identified.

One can express this difficulty of confronting expectation and observation in a different way. Observers usually report their results in terms of a luminosity function, e. g., dN/dM_V. As noted above, this is not a directly measured quantity, but suppose it were. What do we have to know to convert a luminosity function into the theoretically desired mass function? The relation between the luminosity function and the mass function is given by the chain rule:

$$dN/dM_V \;=\; (dN/dm) \times (dm/dL) \times dL/dM_V. \qquad (4)$$

Here N is the measured number of stars, m is the mass, L the total luminosity, V is the observing band, and M_V is the absolute magnitude. For any color, (dm/dL) must be very close to zero near the transition region between nuclear burning (M dwarfs) and gravitationally supported objects (brown dwarfs). The luminosity changes by more than two orders of magnitude while the mass changes by less than 10 %. Thus we have to divide by a very tiny quantity, (dm/dL), in order to convert the measured values into a mass function. What's more, (dm/dL) is not very well determined just around the transition region where, among other things, the age of brown dwarfs strongly affects their luminosity. For simplicity, I have suppresed the age dependence in equation (4). But, the biggest uncertainty in determining (dN/dm) comes from the last term in equation (4), the color relation. Theorists and observers alike agree they don't know how to determine reliably (dL/dM_V).

I conclude that it will not be easy to infer a mass function of brown dwarfs from observations obtained with current techniques.

My impression is that the theory is in good shape for the questions that the theorists consider themselves responsible for answering. The mass-luminosity-age relation, at least asymptotically, is well specified and the uncertainties in the physics of the interior are not appreciable (compared especially to the uncertainties resulting from the atmospheric opacities). I am especially impressed by the relative unanimity regarding the mass estimates for VB8 B based upon the available observational information. The theorists at this conference all seem to agree that the best estimate lies in the range:

$$M = (0.06 \pm 0.02)M_\odot. \qquad (5)$$

Detailed studies of the kinematics of samples of stars moving perpendicular to the galactic disk require that about half of the matter in the solar vicinity be in a form that has not yet been observed. Thus

$$\rho_{\text{unobserved}} = (0.10 \pm 0.05)M_\odot \text{pc}^{-3}. \qquad (6)$$

I think that the most plausible form in which this unseen matter could reside is faint stars. This is the most conservative solution, since we do observe stars with a total local mass density that is within a factor of three of what is required to explain the disk missing matter. Moreover, it is perfectly plausible that the stellar mass function is everywhere continuous but that most of the stars around today happen to lie below the minimum mass for hydrogen burning (equation 2). We would be lucky if it were otherwise but we have no justification for imposing our good fortune as a necessary condition for an acceptable description of the observations. All of the available observations are consistent with the conservative interpretation that the missing disk mass is mostly in brown dwarfs. The trunover of the observed luminosity function in the range of absolute visual magnitudes between 12-14 may simply reflect the steepening of the bolometric correction in this region. In fact, detailed models discussed at this conference predict just that effect. Finally, if we combine the analysis of the z-motions of nearby stars with the measured rotation velocity at the solar position we are forced to conclude that the exponential scale height of the unseen disk material must be less than 0.7 kpc. This constraint is easily satisfied if the local unseen material is in the form of faint stars, but is not easy to understand if the material were in the form of (dissipationless) elementary particles.

HOW TO SUCCEED AT OBSERVATIONAL ASTRONOMY

I think that the unseen disk matter is probably in the form of stars that are not massive enough to burn hydrogen, i. e., brown dwarfs. Their total local mass density is given by equation (6). There should be many of these small mass stars within the near vicinity of the Sun. But, it may be difficult to detect them (because they are faint) and to count their number (because we don't know how to relate their to-be-observed spectra to their masses). The major common task we face is to bring theory and observation closer together so that we can go from a luminosity function to a mass function (see the discussion of equation 4).

Let me close by making some comparisons between our field, the study of brown dwarfs, and two other astronomical fields: black holes and quasars.

From a theoretical point of view, the basic aspects of black holes have been relatively well understood for almost half a century. There have been many observational searches for black holes that were motivated by this brilliant theoretical work. Nevertheless, relatively few certain identifications have been made, at most a few. In brief, for black holes the record shows: **much theory, few objects**.

On the other hand, quasars were discovered serendipitously, by their radio emission (which is not even a prominent characteristic of most quasars). Quasars were certaintly not predicted. When they were discovered, they were unanticipated, widely misunderstood, and (by some) unwanted. Nevertheless, many (i.e., thousands) quasars are now definitely identified. In summary, the record shows for quasars: **uncertain theory, many objects.** A rather similar history, at least regarding their discovery, applies to infrared sources, x-ray sources, pulsars, and γ-ray bursts.

What does this brief historical summary teach us? *How can one succeed at observational astronomy?* It seems to me that recent history teaches us one clear lesson: The best way to succed at observational astronomy is to avoid theoreticians like the plague!

This theoretical work was supported in part by the National Science Foundation grants PHY-8217352 and by NAS8-32902.

INDEX OF NAMES

INDEX OF STARS AND BROWN DWARFS

INDEX OF SUBJECTS